STUDY GUIDE AND PROBLEM SOLUTIONS

To Accompany

BUSINESS STATISTICS: A DECISION-MAKING APPROACH

Prepared By:

V. Lyman Gallup

David F. Groebner

Patrick W. Shannon

Copyright © 1981, by Bell & Howell Company
All Rights Reserved.

Charles E. Merrill Publishing Company
A Bell & Howell Company
Columbus, Ohio 43216

MESSAGE TO THE STUDENT

If you are like many students we have had in our Introductory Statistics classes, you are probably more than a little bit apprehensive about this course. Unfortunately, Statistics is a subject which, for too many people, brings out the worst in them academically. Students who swear they have never had trouble with any other course often do poorly in Statistics.

We think a basic reason for the bad reputation Statistics usually has and the associated poor performance by many students is related to the way many of you attempt to "learn" the material and prepare for exams. All too often, we find that students attempt to memorize terms, equations, and even entire problem solutions. While memory is an important ingredient in the learning process, it does not substitute for <u>understanding</u>.

We are convinced that if you acquire an understanding of the logic of Statistics, you will have little problem with this course. To help you better understand, we have developed this <u>Study Guide And Problem Solutions</u> manual for your use. Rather than offer you a fill-in-the-blank format like most other study guides, we have chosen to provide you with a complete set of class notes for each chapter in the text. These notes highlight the most important concepts in each chapter and provide you with supplemental explanations of the particularly difficult subjects.

The notes have been organized to correspond directly to the sections in the text. We suggest you use these notes to provide a review of each chapter before working the homework problems and several times before your exams.

We have also provided you with complete solutions to most of the odd-numbered student problems in the text. These worked-out solutions provide the back-up for the answers that are present at the end of the text. They demonstrate our approach to solving the student problems. Through observation of our methods, you should be able to more fully <u>understand</u> the logic of Statistics and recognize why we say "Statistics is a partner in decision making."

TABLE OF CONTENTS

Subject	Page
Chapter 1: The Role Of Statistics In Decision Making	1
1-1: What Does Statistics Involve?	1
1-2: Statistics Is An Aid To Decision Making	1
1-3: Using Statistics In The Problem Solving Process	1
1-4: Is Statistics A Math Course?	2
Solutions To Selected Problems	2
Chapter 2: Data Collection	6
2-1: Levels Of Data	6
2-2: Populations And Samples	7
2-3: Sources Of Data	7
Solutions To Selected Problems	8
Chapter 3: Organizing And Presenting Data	9
3-1: Methods Of Presenting Data	9
3-2: Presenting Data	9
Solutions To Selected Problems	10
Chapter 4: Measures Of Location And Spread	13
4-1: Measures Of Location	13
4-2: Measures of Spread	14
4-3: Coefficient Of Variation	16
Solutions To Selected Problems	17

Chapter 5:	Probability Concepts	21
	5-1: What Is Probability?	21
	5-2: Methods Of Assigning Probability	22
	5-3: Probability Rules	23
	5-4: Bayes' Rule	28
	5-5: Counting Techniques	28
	Solutions To Selected Problems	30
Chapter 6:	Discrete Probability Distributions	40
	6-1: Discrete Random Variables	40
	6-2: Discrete Probability Distributions	40
	6-3: Mean And Standard Deviation Of A Discrete Probability Distribution	41
	6-4: Characteristics Of The Binomial Distribution	42
	6-5: Developing A Binomial Probability Distribution	42
	6-6: Using The Binomial Distribution Tables	42
	6-7: Mean And Standard Deviation Of The Binomial	43
	6-8: Some Comments About The Binomial Distribution	44
	6-9: Poisson Probability Distribution	44
	6-10: Variance And Standard Deviation Of The Poisson	45
	Solutions To Selected Problems	46
Chapter 7:	Continuous Probability Distributions	50
	7-1: Continuous Random Variables	50
	7-2: Continous Probability Distributions	50
	7-3: Characteristics Of The Normal Distribution	50
	7-4: Finding Probabilities From A Standard Normal Distribution	51
	7-5: Other Applications Of The Normal Distribution	55

	7-6:	Normal Approximation To The Binomial Distribution	56
		Solutions To Selected Problems	57
Chapter 8:		Sampling Techniques	72
	8-1:	Reasons For Sampling	72
	8-2:	When To Use A Census	73
	8-3:	Fundamental Sampling Techniques	73
		Solutions To Selected Problems	75
Chapter 9:		Sampling Distribution Of $\overline{X}$	78
	9-1:	Relationship Between Sample Data And Population Values	78
	9-2:	Sampling Error	78
	9-3:	Sampling Distribution Of $\overline{X}$	79
	9-4:	Sampling From Normal Distributions	80
	9-5:	Sampling From Non-normal Populations	80
	9-6:	Finite Correction Factor	81
	9-7:	Decision Making And The Sampling Distribution Of $\overline{X}$	81
		Solutions To Selected Problems	83
Chapter 10:		Statistical Estimation--Large Samples	93
	10-1:	The Need For Statistical Estimation	93
	10-2:	Point Estimation	93
	10-3:	Confidence Interval Estimation	94
	10-4:	Confidence Interval Estimation Of μ_x--Large Samples σ_x Known	94
	10-5:	Confidence Interval Estimation Of μ_x--Large Samples σ_x Unknown	96
	10-6:	Confidence Interval Estimation Of A Population Proportion	96

	10-7:	Confidence Intervals For Estimating The Difference Between Two Population Parameters--Large Samples	97
	10-8:	Conclusions	99
		Solutions To Selected Problems	99
Chapter 11:	Statistical Estimation--Small Samples		109
	11-1:	Small Sample Estimation	109
	11-2:	The Student t Distribution	110
	11-3:	Estimating The Population Mean--Small Samples	110
	11-4:	Estimating The Difference Between Two Population Means--Small Samples	111
	11-5:	Estimating The Difference Between Two Population Means--Paired Samples	112
	11-6:	Effect Of Violating The Assumptions For Small Sample Estimation	113
Chapter 12:	Introduction To Hypothesis Testing		122
	12-1:	Reasons For Testing Hypotheses	122
	12-2:	The Hypothesis Testing Process	122
	12-3:	One-Tailed Hypothesis Tests	124
	12-4:	Two-Tailed Hypothesis Tests	124
	12-5:	Hypothesis Testing--Population Standard Deviations Unknown--Large Samples	124
	12-6:	Type II Errors And The Power Of A Hypothesis Test	125
		Solutions To Selected Problems	126
Chapter 13:	Additional Topics In Hypothesis Testing		137
	13-1:	Controlling Type I And Type II Errors	137
	13-2:	Hypothesis Tests About A Population Proportion	138

13-3:	Hypothesis Testing About The Difference Between Two Population Means	138
13-4:	Hypothesis Testing About The Difference Between Two Population Proportions	139
13-5:	Hypothesis Testing With Small Samples-- σ_x Unknown	140
13-6:	An Alternative Way To Test Hypotheses	140
13-7:	Some Other Hypothesis Tests	141
	Solutions To Selected Problems	143

Chapter 14: Analysis Of Variance — 160

14-1:	Analysis Of Variance--One Way Design	160
14-2:	Tukey's Method Of Multiple Comparison	161
14-3:	Scheffe's Method Of Multiple Comparison	161
14-4:	Other Analysis Of Variance Designs	162
	Solutions To Selected Problems	162

Chapter 15: Hypothesis Testing Using Nonparametric Statistics — 175

15-1:	Chi-Square Goodness Of Fit Test	175
15-2:	Chi-Square Goodness Of Fit Limitations	176
15-3:	Mann-Whitney U Test	176
15-4:	Kolmogorov-Smirnov Two-Sample Test-- Large Samples	177
15-5:	Contingency Analysis	178
15-6:	Kruskal-Wallis One Way Analysis of Variance	179
15-7:	Conclusions	180
	Solutions To Selected Problems	180

Chapter 16: Simple Linear Regression And Correlation Analysis — 198

16-1:	Statistical Relationships Between Two Variables	198

16-2:	Correlation Analysis	199
16-3:	Simple Coefficient Of Determination	200
16-4:	Simple Linear Regression Analysis	200
16-5:	Estimating The Simple Regression Model--The Least Squares Approach	201
16-6:	Significance Tests In Regression Analysis	202
16-7:	Regression Analysis For Prediction	203
16-8:	Regression Analysis For Description	204
16-9:	Regression Analysis For Control	204
16-10:	Conclusions	205
	Solutions To Selected Problems	205
Chapter 17:	Introduction To Multiple Regression	220
17-1:	A Non-quantitative Analogy For Multiple Regression Analysis	220
17-2:	The Multiple Regression Model	220
17-3:	Applying The Regression Model	221
17-4:	Dummy Variables In A Regression Model	223
17-5:	Stepwise Regression Models	224
17-6:	Conclusions	225
	Solutions To Selected Problems	225
Chapter 18:	Time Series Analysis	237
18-1:	Time Series Components	237
18-2:	Analyzing The Variability Of Past Data	238
18-3:	Analyzing The Trend Component	239
18-4:	Analyzing The Seasonal Component	239
18-5:	Analyzing The Cyclical Component	240
18-6:	Analyzing The Irregular Component	240

18-7:	Index Numbers	241
18-8:	Commonly Used Index Numbers	241
18-9:	Forecasting	241
18-10:	Conclusions	243
	Solutions To Selected Problems	243

Chapter 19: Indroduction To Decision Analysis — 253

19-1:	Decision Making Under Certainty	253
19-2:	Decision Making Under Risk	253
19-3:	Decision Making Under Uncertainty	254
19-4:	Subjective Probability Assessment	254
19-5:	Decision Making Criteria	254
19-6:	Decision Tree Analysis	255
19-7:	Conclusions	256
	Solutions To Selected Problems	256

Chapter 20: Bayesian Posterior Analysis — 264

20-1:	Bayes' Rule Revisited	264
20-2:	Bayesian Posterior Analysis	265
*20-3:	Bayesian Posterior Analysis With Subjective Conditionals	
20-4:	The Value Of Information	266
20-5:	Conclusions	267
	Solutions To Selected Problems	

* This material covered in text only

CHAPTER 1

THE ROLE OF STATISTICS IN DECISION MAKING

Statistical techniques are simply tools to be used by decision makers. The statistical techniques do not make decisions, but they do help the decision maker analyze the available information, which should lead to improved decisions.

1-1 What Does Statistics Involve?

Statistics is a collection of techniques used to describe data and to make inferences about a large group of data, based upon only a part of those data.

The procedures discussed in this text are designed to assist the decision maker to better describe his or her data and to improve the quality of the inferences made in such a manner so as to make the decision process effective and efficient.

1-2 Statistics Is An Aid to Decision Making:

Managerial problem-solving can almost always be enhanced by applying a consistent, logical process. This process should include at least these five steps:
1. Define the Problem
2. Search for Likely Causes
3. List Alternative Solutions
4. Evaluate the Solutions
5. Choose a Course of Action and the Measurement Criteria

The failure to adequately consider these five steps almost always dooms the decision maker to sub-optimal decisions and outcomes.

1-3 Using Statistics in the Problem Solving Process:

The application of statistical techniques reinforces and enhances a consistent, logical problem-solving process. If statistical techniques are to be applied, the problem must be defined clearly enough to allow the selection of appropriate tools. In addition, statistical tools can be used to distinguish between plausible causes and likely causes.

Statistical techniques are fundamental to summarizing and presenting data in a useful form.

1-4 Is Statistics a Math Course?

The statistical techniques presented in this text require an understanding of only the fundamentals of algebra. Since statistical tools are often used in analyzing large amounts of data, the necessary tasks can sometimes appear voluminous. However, the mathematical concepts required are really no more complicated than addition, subtraction, multiplication, and division.

Learning the terminology of statistics and the notation is imperative if you are to <u>understand</u> Statistics.

```
***************************************************
*                                                 *
*            SOLUTIONS TO SELECTED PROBLEMS       *
*                                                 *
***************************************************
```

1-7

$$X = 2 - 3((2)(2.6) - (-1.7)(-3.8)) + 6(2.6)(-3.8) - (-3.8)^2$$

$$X = 2 - 3(5.2 - 6.46) + -59.28 - 14.44$$

$$X = 2 - 3.78 + 59.28 - 14.44$$

$$X = 43.06$$

1-9

(a) $17.87\% = .1787$
(b) $316\% = 3.16$
(c) $.05\% = .0005$
(d) $20\% = .20$
(e) $1.6783\% = .016783$

1-11

(a) $7X - 6 = 6 - 3X$
solve for X:
$$7X = 6 - 3X + 6$$
$$7X + 3X = 12$$
$$10X = 12$$
$$X = 12/10$$

(b) $5Y - 3(Y-6) = 3(4-2Y) + 6 - 4$
solve for Y:
$$5Y - 3Y + 18 = 12 - 6Y + 2$$
$$2Y + 18 = 14 - 6Y$$
$$2Y = -4 - 6Y$$
$$8Y = -4$$
$$Y = -4/8$$

(c) $4 - 3(Y-6) = 2(X+Y) - 3(X-4)$
solve for X:
$$4 - 3Y + 18 = 2X + 2Y - 3X + 12$$
$$3Y + 22 - 12 = 2Y - X$$
$$5Y + 10 = X$$

(d) $4(\frac{1}{Y} - 6X) + 2X^4 - X^2 = 0$
solve for Y:
$$\frac{4}{Y} - 24X + 2X^4 - X^2 + 0$$
$$\frac{4}{Y} = 24X - 2X^4 + X^2$$
$$\frac{1}{Y} = \frac{24X - 2X^4 + X^2}{4}$$

invert both sides:
$$Y = \frac{4}{24X - 2X^4 + X^2}$$

(e) $A(Z-X^2) + 4YX = -6 + 4(A-Z)$
solve for Z:

$$AZ - AX^2 + 4YX = -6 + 4A - 4Z$$

$$AZ - AX^2 + 4YX + 6 - 4A = -4Z$$

$$-AX^2 + 4YX + 6 - 4A = 4Z - AZ$$

$$-AX^2 + 4YX + 6 - 4A = Z(4-A)$$

$$\frac{-AX^2 + 4YX + 6 - 4A}{4-A} = Z$$

1-13

(a) $\sum_{i=1}^{4} X_i Y_i = (4)(7) + (5)(8) + (9)(10) + (2)(3)$

$\qquad = 28 \quad + \quad 40 \quad + \quad 90 \quad + \quad 6$

$\qquad = 164$

(b) $\sum_{i=1}^{4} X_i \sum_{i=1}^{4} Y_i = (4 + 5 + 9 + 2)(7 + 8 + 10 + 3)$

$\qquad = (20)(28)$

$\qquad = 560$

(c) $\sum_{i=1}^{4} (X_i - \bar{X})(Y_i - \bar{Y}) = 24$

First: $\bar{X} = \frac{(4 + 5 + 9 + 2)}{4} = \frac{20}{4} = 5$

$\bar{Y} = \frac{(7 + 8 + 10 + 3)}{4} = \frac{28}{4} = 7$

Then: $(4-5)(7-7) = 0$
$+ (5-5)(8-7) = 0$
$+ (9-5)(10-7) = 12$
$+ \underline{(2-5)(3-7) = 12}$

$\qquad$ Sum $\quad = 24$

(d) Solve for r:

$$r = \frac{4(164) - 560}{\sqrt{\left[4\sum_{i=1}^{4} X_i^2 - \left(\sum_{i=1}^{4} X_i\right)^2\right]\left[4\sum_{i=1}^{4} Y_i^2 - \left(\sum_{i=1}^{4} Y_i\right)^2\right]}}$$

To find the denominator, set up the following table:

i	X_i	Y_i	X_i^2	X_i^2
1	4	7	16	49
2	5	8	25	64
3	9	10	81	100
4	2	3	4	9
	20	28	126	222

Therefore:

$$r = \frac{4(164) - 560}{\sqrt{[4(126) - 20^2][4(222) - 28^2]}}$$

$$r = \frac{96}{\sqrt{(104)(104)}}$$

$$r = .9230$$

1-15

$$\sum_{i=1}^{n}(X_i - \overline{X}) = \sum_{i=1}^{n} X_i - \overline{X}$$

$$= \sum_{i=1}^{n} X_i - \frac{\sum_{i=1}^{n}\sum_{i=1}^{n} X_i}{n}$$

$$= \sum_{i=1}^{n} X_i - \frac{n \sum_{i=1}^{n} X_i}{n}$$

$$= \sum_{i=1}^{n} X_i - \sum_{i=1}^{n} X_i$$

$$= 0$$

1-17

$$\sum_{i=1}^{n}(X_i - \overline{X})(Y_i - \overline{Y}) = \sum_{i=1}^{n}[X_i Y_i - X_i \overline{Y} - Y_i \overline{X} + \overline{X}\,\overline{Y}]$$

$$= \sum_{i=1}^{n} X_i Y_i - \sum_{i=1}^{n} X_i \overline{Y} - \sum_{i=1}^{n} Y_i \overline{X} + \sum_{i=1}^{n} \overline{X}\,\overline{Y}$$

$$= \sum_{i=1}^{n} X_i Y_i - \frac{\sum_{i=1}^{n} X_i \sum_{i=1}^{n} Y_i}{n} - \frac{\sum_{i=1}^{n} Y_i \sum_{i=1}^{n} X_i}{n} + \frac{\sum_{i=1}^{n} X_i \sum_{i=1}^{n} Y_i}{n\cdot n}$$

combine terms:

$$= \sum_{i=1}^{n} X_i Y_i - \frac{\sum_{i=1}^{n} X_i \sum_{i=1}^{n} Y_i}{n}$$

CHAPTER 2

DATA COLLECTION

An integral step in the decision process is collecting relevant data. Therefore, decision makers, as users of statistical tools, must have a clear understanding of the concepts and terminology which are fundamental to good data collection.

2-1 <u>Levels of Data</u>:

Four levels of data measurement are described in this chapter:

1. Nominal Measurement
2. Ordinal Measurement
3. Interval Measurement
4. Ratio Measurement

The <u>nominal</u> measurement scale requires that we can separate data into groups or classes and then refer to the group by a name or number. We may only conclude that the data points are the same or they are different.
<u>Example</u>: Classification of automobiles owned by people in a certain market area in California as: Sub-compact, Compact, Mid-size, Full Size, and Luxury.

The <u>ordinal</u> measurement scale includes the properties of the nominal scale and in addition, allows for unique ordering of the classes or groups. Ordinal measurement permits comparison so that preference may be established. However, we cannot quantify the magnitude by which one group or class is preferred over another.
<u>Example</u>: The personnel manager who is faced with cutting back on employees in one division of her company might classify the employees as follows: High Level Retention Priority, Medium Level Retention Priority, and Low Level Retention Priority.

The <u>interval</u> scale measurement possesses the measurement properties of the ordinal scale, plus allowing us to quantify the magnitude of the difference between any two data points. The unit of measurement between points remains constant regardless of their position on the scale.
<u>Example</u>: Fahrenheit and Celsius temperature scales.

Ratio scale measurement possesses the properties of interval scales while having a unique zero point. Two ratio measurement scales may have different units of measurement, but they must have the same zero point.

Example: Units of weight such as pounds and grams. If an item weighs nothing, then it weighs 0.0 grams and 0.0 pounds.

2-2 Populations and Samples:

One step toward assuring that a problem is clearly defined requires identifying the population of interest.

A population is a collection of all items of interest to a decision maker. While a population might be a collection of people, it can just as easily be a collection of objects. Examples of populations are: the accounts receivable balances for a department store, the cattle in a feedlot, the spark plugs in a car dealer's parts department, etc....

A sample is a subset of the population. The information in a sample is often used to make inferences about the characteristics of the population from which it was selected.

A census is the measurement of the population. While a census is sometimes possible, in many cases it would be too expensive and too time consuming. In some cases, the measurement requires destroying the product, so a census would be impractical.

Because of the problems involved with taking a census, we rely on sampling and statistical techniques to provide us with inferences about a population and a measure of the quality of that inference.

2-3 Sources of Data:

A large amount of information relevant to the decision maker is generated, compiled, and stored within the corporation or organization. Accounting data, financial records, and operating reports kept by the company can all provide a wealth of information for use in the decision process. The decision maker must make sure that internally-generated data is in a proper format before it can be useful as information.

Data collected within a company or organization is said to come from internal sources.

Data can also be collected from external sources. Government publications are the largest single source of external data.

Before relying on externally published data, a decision maker must make certain that it is a form appropriate to the intended use.

Data collected by someone else for whatever reason becomes secondary data when a second individual uses the data.

Often a decision maker must collect data from <u>primary sources</u> rather than from previously compiled secondary sources. Observation, personal interviews, telephone interviews, and mail questionnaires are popular methods of primary data collection. The method used to collect the data will likely depend upon the type of information the decision maker desires. Failure to properly plan the data collection phase usually leaves the decision maker with abundant data, but little information.

****
SOLUTIONS TO SELECTED PROBLEMS
****

2-3

There are always problems in comparing data gathered from different time periods or from different areas of the country. These problems are often caused by the many differences in social and environmental factors.

Typical differences pertaining to education data might be: differences in course requirements, differences in credit hours required in each academic area, differences in teaching methods, and differences in course content, to name just a few.

2-5

(a) Interval data, since there is no true zero point.

(b) Ordinal measurement, since there is an ordering to the ratings assigned by the drivers. However, it is doubtful that there is a constant and definable measure of distance between the rating classes.

(c) Possible problems:

sound measuring device may be inaccurate or inconsistent;
outside noise levels may be inconsistent;
drivers may have bias;
cars used in the test may not be representative of all cars the company makes

CHAPTER 3

ORGANIZING AND PRESENTING DATA

Data is of little value to a decision maker unless the information which is present in the data can be easily extracted. Effectively organizing and presenting data helps extract the relevant information for use by the decision maker.

3-1 Methods of Presenting Data:

A frequency distribution can be used to summarize and present data so the information present is more easily communicated.

Generally, a frequency distribution will have from 5 to 20 classes. Too few classes will tend to obscure detail in the data, while too many classes fail to allow for adequate summarization.

The classes should be non-overlapping, so that the class membership of an item is unambiguous. Each item in the data set must belong to one, and only one class.

The classes should all have the same class width, if at all possible, to facilitate analysis. Having classes of equal width helps to insure that the data is not misrepresented. However, sometimes a data set will contain extreme values, and equal size classes will not be practical.

The midpoint of a class is halfway between the class limits and is often used to represent the items within the class. Therefore, you should try to establish classes so that the items within each class are evenly spread throughout the class.

A frequency distribution can easily be transformed into a relative frequency distribution. To do this, divide the frequency of each class by the total number of items in the data set. Note that the total number of items is the sum of the individual class frequencies.

3-2 Presenting Data:

Graphical techniques used to present data have the primary goal of making the information in the data more readily apparent. The appropriate graphical technique for displaying data depends largely on the data itself.

Pie charts are often used to illustrate how a total, usually a budget or expenditure amount, has been used or divided.

Histograms are used to present information from a frequency distribution in graphical form. The area of the rectangle in a histogram corresponds to the relative frequency or actual frequency of that class in the frequency distribution. Make certain you label the axes clearly, so that no confusion arises from the presentation, especially if not all classes have the same width.

Bar charts are often used as an alternative to a histogram if a particular point needs emphasis. For many situations, histograms and bar charts are very similar, both in their construction and in their effect.

Cumulative frequency distributions present the data of a frequency distribution, so that the number (or percentage) of items, either less than or greater than a selected item, can be readily determined. Cumulative frequency distributions are constructed by summing the frequencies or relative frequencies.

****
SOLUTIONS TO SELECTED PROBLEMS
****

3-1

(a) Class limits are values which identify the upper and lower limits of each data class.

(b) A histogram is a graph which shows the number of observations falling into each class of a frequency distribution. Generally, the frequency of observations falling into a particular class is represented by the area of the rectangle on the histogram.

(c) A relative frequency distribution is formed by finding the percentage of observations falling into each class interval. The sum of the relative frequencies sums to 1.0.

(d) The cumulative frequency distribution is formed by finding the total number of observations with values less than or equal to a specified class limit.

3-3 The steps involved in developing a frequency distribution are:

1. Gather the appropriate data
2. Decide on the number of classes
3. Determine the appropriate class width, making certain that all observations fall into one and only one class

4. Count the number of cases which fall in each class

3-7

The classes determined in problem 3-6 were based on an interval width of 2500, with a lower limit of the first class at 1000.

Classes	Midpoints
1000 and less than 3500	2250
3500 and less than 6000	4750
6000 and less than 8500	7250
8500 and less than 11000	9750
11000 and less than 13500	11250
13500 and less than 16000	14750
16000 and less than 18500	17250
18500 and less than 21000	19750
21000 and less than 23500	22250
23500 and less than 26000	24750

3-9

The cumulative frequency histogram is developed as follows:

Classes	Cumulative Frequency	
less than 3500	40	
less than 6000	120	(40 + 80)
less than 8500	240	(120 + 120)
less than 11000	490	(240 + 250)
less than 13500	990	(490 + 500)
less than 16000	1690	(990 + 700)
less than 18500	2190	(1690 + 500)
less than 21000	2590	(2190 + 400)
less than 23500	2790	(2590 + 200)
less than 26000	2800	(2790 + 10)

3-11

The cumulative relative frequency distribution is determined by finding the percentage of observations which fall in each class and all preceeding classes. We can simply take the cumulative frequencies determined in problem 3-9 and divide these by the total observations, 2800.

Class		Cumulative Relative Frequency
less than 3500	------------	40/2800 = .01429
less than 6000	------------	120/2800 = .04286
less than 8500	------------	240/2800 = .08571
less than 11000	------------	490/2800 = .17500
less than 13500	------------	990/2800 = .35357
less than 16000	------------	1690/2800 = .60357
less than 18500	------------	2190/2800 = .78214
less than 21000	------------	2590/2800 = .92500
less than 23500	------------	2790/2800 = .99643
less than 26000	------------	2800/2800 = 1.00000

3-13

Your answer may vary slightly, assuming you selected different classes than we selected in problem 3-12. Remember, there is no one right way to establish a frequency distribution. The number of classes and class width is up to the decision maker. Given our approach in problem 3-12, the following represents the appropriate frequency distribution:

Classes	Tally	Frequency
0 - 19.99	‖‖‖ ‖‖‖ ‖‖‖	15
20.00 - 39.99	‖‖‖ ‖‖‖ 1	11
40.00 - 59.99	‖‖‖ 111	8
60.00 - 79.99	‖‖‖ 1111	9
80.00 - 99.99	‖‖‖ 111	8
100.00 - 119.99	1111	4
120.00 - 139.99	11	2
140.00 - 159.99	111	3
	Total	60

3-15

Your report should mention the <u>skewed</u> nature of the distribution. This may reflect the type of customer the store charges to. The store may desire to have the accounts receivable balances small, for the most part. This would tend to help the store's cash flow position.

CHAPTER 4

MEASURES OF LOCATION AND SPREAD

Statistical tools which help summarize the information in a data set and facilitate the communication of that information are a necessity for the decision maker. While the graphical techniques presented in Chapter 3 can often satisfy this need, in many instances a numerical measure of the data is more appropriate. Numerical descriptive techniques provide communication about the information in a data set without the need for graphs and pictures. These numerical measures can be easily used to compare data sets. Numerical descriptive measures are usually classified according to whether they measure the location (central tendency) or the spread (dispersion) of a data set.

4-1 Measures of Location:

The mode is defined as the observation found most frequently in a data set. The mode is the appropriate measure of central tendency in those situations for which a typical size or most popular style must be selected.

The median is the middle observation in a set of data which has been ordered by magnitude. The median is a useful measure of central tendency if the data contain extreme observations. Income distributions are often characterized by many low- and middle-level income and a few large incomes. The use of the median to measure the center or location of these income data prevents the extreme observations from being given more weight than any other income level.

If the data are presented in a frequency distribution, the median class can easily be identified as the class containing the middle value. However, the true median can only be estimated. Since the original values are lost when data are grouped, we usually assume the items in the median class are evenly spread between the class limits.

The arithmetic mean is determined by summing all the values in a data set and dividing this sum by the number of items in the data set. The equations for the population mean and the sample mean are given as:

$$\text{Population} \quad \mu_x = \frac{\sum_{i=1}^{n} X_i}{N} \qquad\qquad \text{Sample} \quad \overline{X} = \frac{\sum_{i=1}^{n} X_i}{n}$$

While the method of computation for the population mean and the sample mean is the same, it is important to remember that the symbolic notation differs. The population mean is noted as μ (mu), and the sample means is symbolized by $\bar{X}$.

The mean is a popular measure of central tendency because of its rigidly defined formula, and because each observation in the data set is used in its computation.

If the data have been grouped into a frequency distribution, the mean can only be estimated, because we lose the identity of the individual values. The accuracy of the estimated mean depends on how well the midpoint of each class represents the items in the class. This is illustrated by the equation for the mean for grouped data which weights the frequency in each class by the class midpoint.

$$\text{Grouped data mean} \quad \mu_x \cong \frac{\sum_{i=1}^{c} f_i M_i}{N}$$

Sets of data which contain extreme observations in a particular direction are said to be skewed. The mean is affected by the extreme cases and will be pulled toward the extremes. As previously indicated, when the data are skewed, the median is often preferred over the mean as a measure of location.

In general, it should be remembered that measures of location should be selected to represent the typical nature of the data set. Since the mean, median, and the mode will often have different values, we must be careful about which one of these measures we select to represent the center of our data for use in the decision making process.

4-2 Measures of Spread:

Decision makers must often be concerned with the variation in the data they are using for informational purposes. For example, product-oriented businesses not only desire products which, on the average are of high quality; they also want products which are consistently of high quality. They want little variation.

The simplest measure of variation or spread is the range which is the difference between the largest and the smallest observations in the data. However, since the range is determined by considering only the two extreme values in the set, it usually does not provide much information about the variation for the majority of observations. The size of the range is seriously affected by extreme observations in the data.

In general, decision makers desire a measure of dispersion which can be used in conjuction with a measure of location to effectively describe a set of observations. The standard deviation and the variance are the most appropriate measures of variation in this respect, since they measure dispersion in the data around the mean of the data.

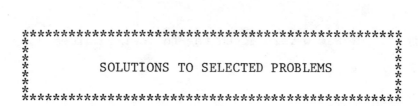

SOLUTIONS TO SELECTED PROBLEMS

4-1

The necessary steps in determining the median are:

(a) Put the data in numerical order.
(b) Find the middle observation. If there are an even number of observations, take the average of the middle two values.

The median is the value above which 50% of the data fall and below which 50% of the data fall when the data have been arranged in numerical order.

4-5

The range is obviously an easier measure to compute. However, the standard deviation has some advantages. First, it incorporates all the data rather than just the two extreme values. It is not as sensitive to the extreme values. The standard deviation also takes into account the mean, and is actually a measure of the dispersion of the data around the mean. Like the range, the standard deviation is measured in the original units.

4-7

As discussed in the text, the mean is sensitive to extreme values and can thus be pulled in one direction or the other. If the data are highly skewed, the median is often preferred. The mode also has some advantages over the mean when we desire to measure the value that is most likely to occur, such as in determining demand for clothing sizes.

If another measure besides the mean is used to measure the center of the population or sample, the standard deviation would not be the desirable measure of dispersion, if we wish to measure the spread around the center. Other measures such as the average absolute median deviation could be employed. We have chosen not to discuss measures of dispersion other than the range and standard deviation in this text. This does not mean that these other measures do not have application. We suggest you consult the chapter references for a discussion of alternative measures of dispersion.

4-9

To respond to this question, we must utilize some of the values determined in problem 4-8. Specifically for Jack Hammer's company:

$\bar{X} = 292.4$ hours
$S_x = 141.179$ hours

Then for I. M. Saw we are given the following values:

$\overline{X}$ = 600 hours

S_x = 50 hours

To determine the relative dispersion in each of the company's labor hours distribution, we should compute the coefficient of variation for each as follows:

Hammer   Saw

$CV = \dfrac{141.179}{292.4}(100.0)$ $CV = \dfrac{50}{600}(100.0)$

CV = 48.29%   CV = 8.33%

Thus, it is clear thet I. M. Saw's distribution of hours is more consistent than Hammer's, since Saw's CV is smaller than Hammer's. Of course our conclusion is based upon a sample which may not represent the population perfectly.

4-11

We begin by referring to problem 10 and constructing a frequency distribution for revenue, assuming the motel collects $20.00 for each room that is occupied.

Revenue Group	M_i Midpoint	f_i Frequency
$0 to $600	300	250
$620 to $1200	910	1200
$1220 to $1800	1510	1900
$1820 to $2400	2110	1100
$2420 to $3000	2710	550
		5000

We find the <u>mean revenue</u> as follows:

$$\overline{X} = \dfrac{\Sigma f_i M_i}{n} = \dfrac{(300)(250) \ldots + (2710)(550)}{5000}$$

$\overline{X}$ = $1569.50

Thus, the average nightly receipts at the motel are $1569.50, as determined from the grouped data shown in problem 10.

To find the standard deviation, we apply the following equation for grouped data:

$$S_x = \sqrt{\frac{f_i(M_i - \overline{X})^2}{n - 1}}$$

Thus,

$$S_x = \frac{250(300 - 1569.50) + \ldots 550(2710 - 1569.50)}{4999}$$

$$S_x = \$627.48$$

Thus, the standard deviation for motel revenue is $627.48.

4-13

Having kept the expense record for a typical week and finding the week's expenses to be $600.00, we would first determine how many standard deviations the $600.00 is from the mean, as determined in problem 4-12.

In problem 4-12 we computed:

$$\overline{X} = \$186.67$$

$$S_x = \$140.14$$

We might let Z equal the number of standard deviations a value is from the mean. Then we can solve for Z as follows:

$$Z = \frac{X - \overline{X}}{S_x}$$

So in this example:

$$Z = \frac{600.0 - 186.67}{140.14} = 2.94$$

Thus, this week's expenses are 2.94 standard deviations above the average. We might conclude that either this was a very unusual week, or that the mean and standard deviation computed from the expense records are questionable. Possibly, the sample data was not representative of all weeks.

4-15

The equation for determining the mean for grouped data assumes that the class midpoints represent all the values in the class. There are two ways of thinking of this. First, we can assume that all values in the class fall at the midpoint, or second, we can assume that the values in the class are evenly spread through the class.

To the extent that our assumptions are true, we will have a good estimate of the mean from the grouped data. Otherwise, our calculation from the grouped data may be quite a bit different from the actual mean.

The same assumptions hold when calculating the standard deviation from grouped data. Again, if we do our job of developing the frequency distribution well, our calculation should be quite close to the standard deviation we would find if we computed it from the raw data.

4-17

The best way to compare the relative variation between two data distributions is to compute the coefficients of variation for each data set and compare these values.

For the height: $CV = \dfrac{S_x}{\overline{X}}(100.0) = \dfrac{2.5}{69.5}(100.0) = 3.60\%$

For weight: $CV = \dfrac{S_x}{\overline{X}}(100.0) = \dfrac{12.0}{177}(100.0) = 6.78\%$

Based on these data, the weight of the passengers has greater variability than height. In fact, on a relative basis, the variability for weight is almost twice that of height.

CHAPTER 5

PROBABILITY CONCEPTS

Probability is a tool for reasoning from the general to the specific. Given the knowledge of the form of a population (the general), the concepts of probability enable us to describe the results we should expect in a random sample (the specific) from the population. This vital link between the population and the sample is the foundation for statistical inference. Decision makers need to understand these probability concepts in order to objectively evaluate their statistical inferences.

5-1 <u>What is Probability</u>?

Three common methods for assigning probabilities are:

1. Relative frequency of occurrence
2. Classical probability
3. Subjective probability

While these three methods differ, the terminology which is fundamental to assigning probabilities is fairly standard. To insure a clear understanding of the concepts of probability, you need to be comfortable with the terminology.

A <u>sample space</u> is the collection of all possible outcomes which can result from an experiment. Each individual outcome of an experiment is called an <u>elementary</u> event. A collection of elementary events is an <u>event</u>.

For example, if we randomly select one card from an ordinary deck of 52 playing cards, the 52 cards collectively compose the <u>sample space</u>. Each card individually is an <u>elementary event</u>. The collection of the 26 red cards is one example of an <u>event</u>.

Two events are <u>mutually exclusive</u> if they have <u>no</u> elementary events in common. Red cards and black cards are mutually exclusive, since a card can't be red and black, simultaneously.

Two events are <u>independent</u>, if the knowledge that one event has (or has not) occurred does not alter the probability that a second event will (or will not) occur. For example, suppose we experiment: draw a card from a well-shuffled deck of playing cards, observe the color (red or black), replace the card and reshuffle, select a second card, and observe the color. Since the experiment is identical for both selections, knowledge of the color of the card selected on the first draw <u>does not</u> change the probability of getting a red or black card on the second draw. The event (red or black) on the first draw is <u>independent</u> of the event (red or black) on the second draw. The kind of sampling described in this experiment is called <u>sampling with replacement</u>.

If two events are not independent, they are said to be <u>dependent events</u>. For example, considering the previous experiment, suppose we change the rules for the second draw to be that the second card is selected <u>without replacing</u> the first card in the deck. Now suppose we know that the first card selected was a red card which had a 26/52 or 1/2 probability of occurring. The probability that the second card selected is red, is now 25/51, which is slightly less than 1/2. Thus, when we sample without <u>replacement</u>, the resulting events are dependent.

5-2 <u>Methods of Assigning Probability</u>:

Recall, there are three methods of assigning probability:

1. Relative frequency of occurrence
2. Subjective probability assessment
3. Classical probability assessment

<u>Relative frequency probabilities</u> are determined by constructing a ratio of the number of times an event was observed to the number of times the experiment was conducted. Then if the experiment has been conducted a large number of times, the relative frequency of occurrence for the event of interest should be very close to the probability of that event occurring.

For example, suppose a factory quality control manager has tested 5000 parts produced by his company's assembly line and found 500 defective parts. Based upon the relative frequency with which defective parts were observed, the quality control manager might assess the probability of a defective part being produced as:

$$P(\text{defective}) = \frac{500}{5000} = .10$$

<u>Subjective probability</u> is a measure of personal uncertainty regarding the occurrence of an event. Often, we cannot collect relative frequency data, and the only means of assessing the probability of an event occurring is to subjectively assess it. This is a common activity for business decision makers who must subjectively assess probabilities about such things as potential sales, competitor prices, inventory lead time, weather, and individual human reactions to various events. These subjective probabilities reflect the decision maker's <u>state of mind</u> regarding the likelihood of the event of interest occurring.

<u>Classical probability</u> assessment requires a knowledge of all the possible outcomes of an experiment and rests on the condition that all possible

<u>elementary events are equally likely</u> to occur. We can easily develop classical probabilities for the card selection example used earlier, since all 52 possible elementary events are known, and if the selection is done randomly, all the elementary events are equally likely to occur.

Often in business, it is difficult to determine all the possible outcomes of an experiment, and usually the outcomes aren't equally likely. Thus, while classical probability is appropriate for most games of chance, it is generally not a feasible method of assessing probabilities for business decision makers. However, classical probability is often the basis for learning the basic rules of probability which apply to all forms of probability assessment.

5-3 <u>Probability Rules</u>:

Listing all the possible outcomes of an experiment becomes impractical, even for some relatively basic problems. Therefore, we need a set of procedures for evaluating the probability of an event or series of events that does not require complete enumeration of all outcomes.

First, we must recognize that probabilities, since they relect the relative likelihood of occurrence, cannot be negative and cannot exceed one. This is reflected in Probability Rule 1:

<u>Probability Rule 1</u>:

$$0.0 - P(E_i) - 1.0$$

and

$$0.0 - P(e_i) - 1.0$$

Further, since some outcome must occur when an experiment is conducted, the sum of the probabilities assigned to the elementary events of an experiment must equal one. This is reflected in Probability Rule 2:

<u>Probability Rule 2</u>:

$$P(e_i) = 1.0$$

If we return to the card selection example, we can define an event E_1: a red card is selected. Since the card selected is either <u>red or not red</u>, we refer to the event, <u>not red</u>, as the <u>complement</u> of the event, <u>red</u>. The complement is labeled E_1. Therefore, the probability of an event plus its complement must equal one, because we have divided the sample sapce into two mutually exclusive events.

<u>Addition Rules</u>:

To find the probability of occurrence of event A <u>or</u> event B, simply add (+) the individual probabilities of event A and event <u>B</u>, less any joint probability of the two events occurring at the same time. If the two events are <u>elementary events</u>, or if they are <u>mutually exclusive events</u>, the probability is found by adding the individual probabilities. These concepts are known as the addition rules and are illustrated in Probability Rules 3, 4, and 4a.

Probability Rule 3:

Addition rule for elementary events:

If $E_i = (e_1, e_2, e_3)$

Then:

$$P(E_i) = P(e_1) + P(e_2) + P(e_3)$$

Probability Rule 4:

Addition rule for any two events, A and B:

$$P(A \text{ or } B) = P(A) + P(B) - P(A \text{ and } B)$$

Probability Rule 4a:

Addition rule fur mutually exclusive events, A and B:

$$P(A \text{ or } B) = P(A) + P(B)$$

Examples:

Recall the card selection example. Since there are 52 cards in the deck, there are 52 elementary events.

Define E_1 = a spade is selceted

E_2 = a heart is selected

E_3 = a seven is selected

Since the selection is assumed to be random, each of the 52 elementary events is equally likely to occur, and by classical probability assessment:

$$P(E_1) = \frac{13 \text{ spades}}{52 \text{ cards}} = 1/4$$

$$P(E_2) = \frac{13 \text{ hearts}}{52 \text{ cards}} = 1/4$$

$$P(E_3) = \frac{4 \text{ sevens}}{52 \text{ cards}} = 1/13$$

Thus, if we wish to find the probability of observing either a heart or a spade, we apply rule 4a, because the events are mutually exclusive:

$$P(E_1 \text{ or } E_2) = P(E_1) + P(E_2)$$
$$= 1/4 + 1/4$$
$$P(E_1 \text{ or } E_2) = 1/2$$

If we wish to find the probability of selecting either a spade or a seven, we use <u>rule 4</u>. Note that these events are not mutually exclusive, because it <u>is possible</u> to get a seven of spades.

$$P(E_1 \text{ or } E_3) = P(E_1) + P(E_3) - P(E_1 \text{ and } E_3)$$
$$= 1/4 + 1/13 - 1/52$$
$$= 13/52 + 4/52 - 1/52$$
$$P(E_1 \text{ or } E_3) = 16/52$$

Note that we have subtracted the probability of selecting a seven of spades ($P(E_1 \text{ and } E_3)$), because it is included in both E_1 and E_3, and we must avoid the double counting.

Conditional Probability Rules:

Many probability questions require that two events occur at the same time. In other cases, additional information is obtained which causes us to revise earlier probability assessments. To illustrate, define events E_1, E_2, and E_3 as before and;

$$E_4 = \text{a black card is selected}$$

Now suppose we know that a black card has been selected, and we wish to determine the probability that the card is also a spade. In probability notation, we seek $P(E_1|E_4)$. To find this probability or other probabilities of this form, we must apply the rule for conditional probability, Probability Rule 5:

Probability Rule 5:

Conditional probability for any two events:

$$P(E_1|E_4) = P(E_1 \text{ and } E_4)/P(E_4)$$

We are given the information that the card is black, which effectively restricts the sample space from the original 52 possible outcomes to the 26 cards which are black. So we find:

$$P(E_1|E_4) = \frac{P(E_1 \text{ and } E_4)}{P(E_4)} = \frac{\text{\# of black spades}}{\text{\# of black cards}}$$

$$P(E_1|E_4) = \frac{13}{26} = 1/2$$

We read $P(E_1|E_4)$ as the probability of E_1, given E_4. This is a conditional probability, because the possible occurrence of E_1 is conditioned on the information that E_4 has already occurred. In the classical probability context, conditional probabilities are the ratio of the number of outcomes common to both events to the number of outcomes associated with the event which is given to have occurred.

Multiplication Rule:

If we begin with Probability Rule 5 for conditional probability:

$$P(E_1|E_4) = \frac{P(E_1 \text{ and } E_4)}{P(E_4)}$$

Then rearranging the above equation, we obtain Probability Rule 6:

Probability Rule 6:

Multiplication rule for two events:

$$P(E_1 \text{ and } E_4) = P(E_4) * P(E_1 | E_4)$$

We can see that Rule 6 provides us with a means of determining the probability of both events E_1 and E_4 occurring simultaneously. Thus, a <u>joint probability</u> is the probability of the joint occurrence of two or more events.

Naturally, we cannot determine both $P(E_1|E_4)$ and $P(E_1 \text{ and } E_4)$, except in classical probability situations where we can count all the sample points. We need two of the three pieces of information in Rule 5 or Rule 6 to find the third.

Independent Events:

Sometimes the fact that one event is given to have occurred does not cause us to change our original probability assessments. For example:

$$P(E_3|E_4) = \frac{P(E_3 \text{ and } E_4)}{P(E_4)} = \frac{\text{\# of black sevens}}{\text{\# of black cards}}$$

$$P(E_3|E_4) = 2/26 = 1/13$$

which is the same as the original $P(E_3) = 1/13$. When this situation occurs, the events are said to be independent, as depicted in Probability Rule 5a:

Probability Rule 5a:

Conditional Probability for Independent Events:

$$P(E_3|E_4) = P(E_3)$$

Thus, when the events are independent, knowledge of the fact that event E_4 has occurred does not change the probability of the event E_3.

If two events are <u>independent</u>, the probability of their joint occurrence is the product of their <u>individual</u> probabilities. This is illustrated as Probability Rule 6a:

Probability Rule 6a:

Multiplication Rule for Independent Events:

$$P(E_3 \text{ and } E_4) = P(E_3) * P(E_4)$$

For example:

$$P(E_3 \text{ and } E_4) = P(E_4)P(E_3|E_4)$$
$$= (1/2)(1/13)$$
$$P(E_3 \text{ and } E_4) = 1/26$$

also:

$$P(E_3 \text{ and } E_4) = P(E_3) P(E_4)$$
$$= (4/52)(1/2)$$
$$P(E_3 \text{ and } E_4) = 4/104 = 1/26$$

Thus, events E_3 and E_4 are <u>independent events</u>.

5-4 Bayes' Rule:

Bayes' rule is a procedure for computing conditional probabilities. Actually, it is merely a restatement of Probability Rule 5. Bayes' rule is usually appropriate for events which can be sequenced by some natural ordering. As more information is collected, we revise our earlier probability assessments to reflect both our prior assessment and the effect of the new information. In chapter 20, we will extensively utilize Bayes' rule in decision making under uncertainty.

5-5 Counting Techniques:

Oftentimes, the range of possible outcomes for a sampling experiment is so wide, that counting rules are needed to insure that no outcomes are overlooked. Permutations and combinations are two counting techniques which can be used to enumerate the possible outcomes for particular sampling experiments.

Permutations:

If the order of occurrence is important, the method of permutations will be the appropriate method to use to count the number of outcomes. For example, a committee of Mary, Nancie, Paul, Tom, and Roger wishes to select a chairperson and a secretary. The first person selected will chair the committee, and the second will be the secretary. Therefore, (Mary,Paul) is a different outcome than (Paul,Mary), since the order of selection determines which person fills which position.

Permutations provides the number of different pairs possible, keeping track of the order of occurrence:

$$P_2^5 = \frac{5!}{(5-2)!} = \frac{5!}{3!} = \frac{5 \times 4 \times 3 \times 2 \times 1}{3 \times 2 \times 1} = 20$$

Thus, there are 20 different pairs that could be selected which will lead to different officers on the committee.

The general formula for permutations is:

$$P_r^n = \frac{n!}{(n-r)!}$$

Combinations:

In situations where the order of selection is not considered important, we use the method of combinations to count the number of possible outcomes. For example, consider the committee example again. Suppose we merely wish to select a subcommittee of two members. In this case, the order of selection is not important, since (Mary,Paul) is the same pair as (Paul,Mary). We use combinations to determine the number of different subcommittees of size two as follows:

$$C_2^5 = \frac{5!}{2!(5-2)!} = \frac{5 \times 4 \times 3 \times 2 \times 1}{(2 \times 1)(3 \times 2 \times 1)} = 10 \text{ pairs}$$

Thus, there are 10 different subcommittees that could be formed from a five-member committee.

```
***************************************************
*                                                 *
*            SOLUTIONS TO SELECTED PROBLEMS       *
*                                                 *
***************************************************
```

5-1

(a) This probability could be assessed either by the <u>relative frequency of occurrence</u> method, or by measuring the number of days in the past when it actually rained when the weather conditions are the same as for tomorrow, divided by the total number of days with these same weather conditions.
The method of assessment could be <u>subjective</u>, indicating the degree of belief that it will, in fact, rain tomorrow.

(b) This assessment would be <u>subjective</u>, based upon quantitative and qualitative information about the team and the rest of the league.

(c) This type of assessment would most likely be based upon historical records, and would therefore be based upon the <u>relative frequency of occurrence</u> method.

(d) Most likely, the method of assessment would be subjective, although it would be based upon a market research study, as well as past records of product introduction by this company, if available.

5-3

Given:
$$P(A) = .5 \quad P(B) = .4 \quad P(A \text{ and } B) = .2$$

(a) If two events are mutually exclusive, P(A and B) = 0.0, since both events cannot occur simultaneously. Since P(A and B) = .20, the events are <u>not</u> mutually exclusive.

(b) If two events are independent $P(A \text{ and } B) = P(A) \cdot P(B)$. In this case, since $P(A) \cdot P(B) = .20 = P(A \text{ and } B)$, the events are said to be independent.

5-5

Since we can assume that the chairman will vote to keep the president, we wish to determine the probability of selecting four other members who agree with him. While it would be possible to list the sample space, this solution is suited to the use of combinations as a counting technique.

We have:

        5 loyal (6 including the chairman)

        4 against

Four will be chosen from the nine.

$$P(4 \text{ loyal}) = \frac{\text{number of ways to select 4 loyal}}{\text{number of way to select any 4}}$$

$$\frac{C_4^5}{C_4^9} = \frac{\frac{5!}{4!(5-4)!}}{\frac{9!}{4!(9-4)!}} = \frac{5}{126}$$

5-7

Now, in order for the vote to go 4-1 against the president, all four members chosen for the committee would have to be against the president. Note that we still assume that the president will be in the group. To find this probability, we use combinations.

$$P(4 \text{ against}) = \frac{C_4^4}{C_4^9} = \frac{\frac{4!}{4!(4-4)!}}{126} = \frac{1}{126}$$

The probability of a 4-1 vote against the president by chance is very small.

5-9

The probability of making one sale in two calls, assuming the calls are <u>independent</u> and that the probability of sale on any call is .30, is determined as follows:

$$P(1 \text{ sale in two calls}) = P(S \text{ and } \bar{S}) + P(\bar{S} \text{ and } S)$$

$$= P(S)P(\bar{S}) + P(\bar{S})P(S)$$

$$= (.3)(.7) + (.7)(.3)$$

$$P(\text{a sale in two calls}) = .42$$

5-11

We are asked to determine the probability that Dave can make it through the five decisions without getting fired. We can begin by listing the sample space for the events which satisfy the criteria of one or fewer mistakes.

Let:

    G = good decision   B = bad decision

The sample space is:

    GGGGG = no errors
    GGGGB
    GGGBG
    GGBGG = 1 error or bad decision
    GBGGG
    BGGGG

The probabilities of these outcomes are:

$$P(5 \text{ good}) = (.8)(.8)(.8)(.8)(.8) = .32768$$

and for one of the ways $P(4G, 1B) = (.8)(.8)(.8)(.8)(.2) = .08192$

Since there are <u>five</u> ways we can get one mistake:

$$P(4G, 1B) = 5(.08192) = .4096$$

Thus:

$$P(\text{not fired}) = .32768 + .4096 = .73728$$

Given this high probability of getting fired, it would appear that Dave should not take this particular job.

5-13

Given the probabilities computed in problems 5-11 and 5-12, and the recommended decisions based upon these probabilities, there is no chance that Dave will be offered the job and that he would accept. However, as is always the case, different decision makers might view the probabilities differently and thus arrive at decisions which differ from those we have recommended.

5-15

(a) We can use combinations to assist in solving this probability problem as follows:

$$P(1 \text{ defective}) = \frac{\text{number of ways to get 1 defective}}{\text{number of ways to draw 5 tubes}}$$

$$= \frac{C_1^5 \, C_4^{15}}{C_5^{20}}$$

$$= \frac{\frac{5!}{1!(5-1)!} \frac{15!}{4!(15-4)!}}{\frac{20!}{5!(20-5)!}}$$

$$= \frac{(5)(1365)}{15504}$$

P(1 defective) =

(b) We can either use combinations or listing of the sample space to solve this. Suppose we list the following:

no defective = G G G G G

The probability of this event is:

P(no defective) = (15/20) (14/19) (13/18) (12/17) (11/16)

= .194

(c)
P(2 or fewer defectives) = P(0) + p(1) + P(2)

From parts (a) and (b), we have:

P(0) = ~~1.94~~ 0.194

P(1) = .440

Now we find the chances of finding two defectives as follows:

$$P(2) = \frac{\text{number of ways to get 2 defectives}}{\text{total ways to get 5 tubes}}$$

$$= \frac{C_2^5 \, C_3^{15}}{C_5^{20}}$$

P(2 defectives) = $\frac{4550}{15504}$ = .293

Thus:

P(2 or fewer) = .194 + .440 + .293

= .927

5-17

   We can solve this problem by setting up a <u>joint frequency distribution</u> table:

	Favor	Not Favor	
Production	370	380	750
Office	345	105	450
	715	485	

   We can change the joint frequencies to <u>joint relative frequencies</u> and set up another table as follows:

	Favor	Not Favor	
Production	.308	.317	.625
Office	.288	.087	.375
	.596	.404	

   Then, using the above table and letting the joint relative frequencies represent the probabilities of interest, we can solve for the following:

   (a) P(Favor) = .596

   (b) P(Not Favor and Office) = .087

   (c) For independence to hold, we need:

         P(Not Favor and Office) = P(Not Favor)P(Office)

              .087 $\neq$ (.404)(.375)

Therefore, the attitude on this issue and type of work <u>are not</u> independent.

5-19

   We first construct a <u>joint frequency distribution</u>:

	Increase	No Increase	
Blue Chip	30	20	50
Growth	20	10	30
	50	30	

Next we construct a relative frequency distribution:

	Increase	No Increase	
Blue Chip	.375	.250	.625
Growth	.250	.125	.375
	.625	.375	

From the joint relative frequency table:

P(Blue Chip and No Increase) = .250

5-21

Referring to the joint relative frequency table in the solution to problem 5-19, we find:

$$P(\text{Growth} \mid \text{No Increase}) = \frac{P(\text{Growth and No Increase})}{P(\text{No Increase})}$$

$$= .125/.375$$

$$= .333$$

5-23

Referring to the work done in problem 5-22, we can list the available probabilities:

P(Tennis) = .30

P(Golf) = .70

P(Jones Win Tennis) = .80

P(Jones Win Golf) = .3

Using Bayes' Rule:

$$P(\text{Tennis} \mid \text{Jones Win}) = \frac{P(\text{Tennis and Jones Win})}{P(\text{Jones Win})}$$

where:

P(Jones Win) = P(Tennis)P(Jones Win | Tennis) + P(Golf)P(Jones Win | Golf)

P(Jones Win) = (.30)(.80) + (.70)(.30)

= .45

and:

$$P(\text{Tennis and Jones Win}) = P(\text{Tennis})P(\text{Jones Win} | \text{Tennis})$$

$$= (.30)(.80)$$

$$= .24$$

Thus:
$$P(\text{Tennis} | \text{Jones Win}) = .24/.45$$

$$= .533$$

Given this information, the probability is only slightly in favor of having played tennis.

5-25

We first ask how it will be possible for M-K to win the bid. The answer is:

(M-K Win and Lite not Bid) <u>or</u> (M-K Win and Lite does Bid)

The probabilities of these events are determined as follows:

$$P(\text{M-K Win and Lite Bid}) = P(\text{Lite Bid})P(\text{M-K Win} | \text{Lite Bid})$$

$$= (.75)(.20)$$

$$P(\text{M-K Win and Lite Bid}) = .15$$

and:

$$P(\text{M-K Win and Lite not Bid}) = P(\text{Lite not Bid})P(\text{M-K Win} | \text{Lite Not})$$

$$= (.25)(.50)$$

$$P(\text{M-K Win and Lite not Bid}) = .125$$

Since M-K can win the bid two ways, we must add the two probabilities:

$$P(\text{M-K Win}) = .150 + .125$$

$$= .275$$

5-27

For the present pitcher: $P(\text{Out}) = .70$
$P(\text{No Out}) = .30$

For the relief pitcher: $P(\text{Out} \mid \text{not Best}) = .40$
$P(\text{Out} \mid \text{Best}) = .90$
$P(\text{Best}) = .70$

We need to determine whether we should bring in the relief pitcher. The decision should depend on which pitcher has the highest probability of getting the batter out. Thus, we need to find the probability that the relief pitcher will get the batter out and compare this to the .70 for the present pitcher.

For the relief pitcher:

$$P(\text{Out}) = P(\text{Best})P(\text{Out} \mid \text{Best}) + P(\text{not Best})P(\text{Out} \mid \text{not Best})$$

$$= (.70)(.90) + (.30)(.40)$$

$$P(\text{Out}) = .75$$

Since the relief pitcher has a .75 chance of getting the batter out, all things considered, we should elect to bring in the relief pitcher.

5-29

We need to determine certain probabilities as stated in the problem:

$P(\text{Difficulty}) = .05$

$P(\text{Electrical} \mid \text{Difficulty}) = .50$

$P(\text{Mechanical} \mid \text{Difficulty}) = .50$

$P(\text{Cut} \mid \text{Mechanical}) = .40$

$P(\text{Cut} \mid \text{Electrical}) - .50$

(a) We are asked to determine the probability of a flight getting cut short. We do this as follows:

$P(\text{Cut}) = P(\text{Difficulty and Mechanical and Cut}) + P(\text{Difficulty and Electrical and Cut})$

$= (.05)(.50)(.40) + (.05)(.50)(.50)$

$= \quad .01 \quad + \quad .0125$

$P(\text{Cut}) = .0225$

(b) This is a problem that requires Bayes' Rule as follows:

$$P(\text{Electrical} \mid \text{Cut}) = \frac{P(\text{Electrical and Cut})}{P(\text{Cut})}$$

From part (a), we already know that P(Cut) = .0225

Then:

$$P(\text{Electrical} \mid \text{Cut}) = \frac{(.05)(.50)(.50)}{.0225}$$

$$= .0125/.0225$$

$$= .556$$

5-31

This problem can be solved using combinations:

$$C_3^{10} = \frac{10!}{3!(10-3)!}$$

$$= 120 \text{ ways}$$

Note that this approach assumes that the order is not important. If order of selection is important, we must use permutations--and we get 720 ways.

5-33

We can use combinations to determine the number of teams that could be selected.

$$C_1^6 \text{ and } C_1^5 \text{ and } C_1^3 \text{ and } C_1^8 = 6 \times 5 \times 3 \times 8 = 720 \text{ teams}$$

5-35

Using six numbers on a plate, the possible licenses run from 000000 to 999999. In each spot on the license we have the choice of ten digits. Thus, we have:

10 x 10 x 10 x 10 x 10 x 10 = 1,000,000 possible license plates.

If the state uses three letters and three numbers, they have a choice of 26 letters and 10 digits on each spot on the license plate. Thus, they have:

36 x 36 x 36 x 36 x 36 x 36 = "many"

If we assume that the sequence is 3 letters followed by 3 digits, we arrive at:

26 x 26 x 26 x 10 x 10 x 10 = 17,576,000 plates

If they use two letters followed by four digits, we arrive at:

26 x 26 x 10 x 10 x 10 x 10 = 6,760,000 possible plates.

5-37

$P(A) = .40 \quad P(Bad \mid A) = .15$

$P(B) = .50 \quad P(Bad \mid B) = .10$

$P(C) = .10 \quad P(Bad \mid C) = .25$

Usining Bayes' Rule:

$P(A \mid Bad) = \dfrac{P(A)P(Bad \mid A)}{P(Bad)} = \dfrac{(.40)(.15)}{(.40)(.15)+(.50)(.10)+(.1)(.25)}$

Most likely

$P(A \mid Bad) = .444$

$P(B \mid Bad) = \dfrac{(.50)(.10)}{.135}$

$\quad\quad\quad\quad = .05/.135 = .370$

$P(C \mid Bad) = 1.0 - (.444) + (.370) = .186$

CHAPTER 6

DISCRETE PROBABILITY DISTRIBUTIONS

As we found in Chapter 5, determining the probability of observing a particular sample outcome can require an extensive listing of all the possible outcomes. Fortunately, many of the decision situations in business possess some common characteristics. The fact that these situations possess common characteristics allows us to model the outcomes with well-known probability distributions. The key then becomes recognizing the characteristics which are necessary to employ different probability distributions.

6-1 Discrete Random Variables:

A random variable is simply a rule for assigning numerical values to the outcomes of an experiment. Random variables can be classified as either discrete or continuous, depending on the kind of numerical values they assign to the outcomes.

Discrete random variables possess a set of distinct values which most often arise from a counting process. The number of half gallon milk cartons demanded per day in a supermarket would be a discrete variable.

Continuous random variables possess an infinity of possible values which typically arise through a measurement process. With an accurate enough measurement device, we could always find at least one more value of the variable between any two selected values. The true weight of the milk in cartons designed to hold 64 fluid ounces (a half gallon) would be a continuous variable. Given any two carton weights, it is always possible to find another weight in between. This is not true, for example, if we count the number of cartons demanded since no other possible value exists between, say, 10 and 11 cartons demanded.

6-2 Discrete Probability Distributions:

A probability distribution is a list of all the values of a random variable and their associated probability of occurrence. Each of these probabilities must be nonnegative and no greater than one. Since the values of the random variable must be mutually exclusive and collectively exhaustive, the sum of the probabilities must equal one. For example:

	X = number of season ski tickets demanded per day	P(X)
	0	.1
	1	.4
	2	.3
Note: $0.0 \leq P(X) \leq 1.0$	3	.2
$\Sigma P(X) = 1.0$		$\Sigma = 1.0$

6-3 Mean and Standard Deviation of a Discrete Probability Distribution:

Often we find it necessary to summarize the information in a probability distribution using descriptive measures of <u>location</u> and <u>spread</u>. One measure of the location of a probability distribution is called the <u>expected value</u> and is represented as E(X). The expected value is the long-run average value of the random variable.

The <u>standard deviation</u> measures the dispersion of a probability distribution just as it measured the dispersion of a set of measurements. The formulas to compute the mean and standard deviation of a probability distribution respectively, are:

$$E(X) = \Sigma X \cdot P(X)$$

and

$$\sigma_X = \sqrt{\Sigma (X - E(X))^2 \cdot P(X)}$$

To illustrate, recall the demand for season ski tickets presented in section 6-2:

X	P(X)	X · P(X)	X-E(X)	(X-E(X))²	(X-E(X))² · P(X)
0	.1	.0	-1.6	2.56	.256
1	.4	.4	.6	.36	.144
2	.3	.6	.4	.19	.048
3	.2	.6	1.4	1.96	.392
		1.6 = E(X)			.84

So: $E(X) = \Sigma X \cdot P(X) = 1.6$

$$\sigma_X = \sqrt{\Sigma (X - E(X))^2 \cdot P(X)} = \sqrt{.84} \approx .917$$

Thus, the expected number of ski tickets demanded per day is 1.6 with a standard deviation of .917.

6-4 Characteristics of the Binomial Distribution:

One common discrete probability distribution is the binomial. We can use the binomial probability distribution for any experiment which possesses the following characteristics:

1. Only two possible outcomes exist on each trial. These outcomes can be called "success" and "failure."

2. There are n identical trials or experiments.

3. The outcome (success or failure) of any one trial is independent of the outcome of any other trial.

4. The probability of observing a success, p, is the same for every trial. Therefore, q = 1 - p is the probability of observing a failure and q must also remain constant.

Then, typically, we are interested in the number of successes observed in n trials. However, a success is defined by the decision maker. For example, the inventory control manager might call a missing part from inventory a success if her objective was to determine the number of missing parts.

6-5 <u>Developing a Binomial Probability Distribution</u>:

Because the binomial probability distribution is characterized by a set of special conditions, a general formula has been developed to compute binomial probabilities. The general formula for the probability of X1 successes in n trials is:

$$P(X1) = \frac{N!}{X1! \; X2!} \; p^{X1} \; q^{X2}$$

where:

$X2 = n - X1$ = number of failures
p and q = probability of success and failure, respectively

This formula is appropriate because p and q remain constant, successive trials are independent, and the number of orderings for X1 successes and X2 failures can be evaluated using distinct permutations. Since each different ordering of X1 success and X2 failures has the same probability of occurrence, we can multiply the probability of one representative outcome by the number of distinct permutations. This is all the binomial formula really does.

6-6 <u>Using the Binomial Distribution Tables</u>:

Given unique values for the number of trials, n, and the probability of a success, p, all binomials have the same probability distribution. Therefore, rather than evaluate the probabilities using the binomial formula each time, we can find many common binomial probabilities in a table. Table A in the appendix of the text provides binomial probabilities for a variety of values for n and p.

To use the table, we use the following process:

1. Locate the section of the table corresponding to the appropriate sample size, n.

2. Identify the column corresponding to the selected value of p. If $p \leq .50$, use the entries for number of successes from the left-hand margin. If $p > .50$, use the entries for number of successes from the right-hand margin of the binomial table.

For example, to find the probability of exactly 8 successes in n = 20 trials, if p = .45, we locate n = 20 and p = .45 in the table. Since $p \leq .5$, we use the number of successes recorded in the left-hand margin, and we find $P(X1 = 8) = .1623$.

Recall that "successes" and "failures" for binomial outcomes are labeled at the discretion of the analyst. Since this is true, we could have counted the 20 - 8 = 12 "failures" in the previous example and located the probability of exactly 12 failures in the binomial table. To do this, notice that if the probability of a success is .45, then the probability of a failure must be 1 - .45 = .55. We find q = .55 at the bottom of a column and locate 12 failures in the right-hand margin. Since 8 successes in n = 20 trials with q = .55, it should be no surprise that the probability entered in the body of the table is .1623.

6-7 Mean and Standard Deviation of the Binomial Distribution:

The mean, or expected value of the binomial, depends only on the number of trials, n, and the probability of a success, p. Therefore, to calculate the expected value of a binomial, we use:

$$E[X] = \mu_X = n \cdot p$$

Similarly, the equation for the standard deviation of a discrete probability distribution simplifies to:

$$\sigma_X = \sqrt{n \cdot p \cdot q}$$

if the conditions of the binomial are met.

Thus, for a binomial distribution with n = 100 and p = .50, the mean and standard deviation can be determined as:

$$E[X] = \mu_X = (100)(.5) = 50$$

$$\sigma_X = \sqrt{(100)(.5)(.5)} = \sqrt{25} = 5$$

6-8 Some Comments About the Binomial Distribution:

When the probability of a success equals .5, the probability of a failure must also equal .5. When this condition exists, the binomial probability distribution is symmetric, regardless of the number of trials.

Whenever p differs from .5 in either direction, the binomial probability distribution is not symmetric and is said to be skewed. However, the binomial approaches symmetry for any level of p if the sample size is large enough.

6-8 Poisson Probability Distribution:

In some situations we can count the number of occurrences of interest (like binomial successes), but the number of nonoccurrences (or failures) cannot be counted. We can count the number of phone calls per hour coming into a hotel switchboard, but we cannot count the number of calls that didn't come into the switchboard. Since we cannot determine either the number of failures or the total number of trials in cases like this one, the binomial probability distribution is not appropriate.

The Poisson probability distribution provides a good model for processes which generate rare physical events randomly and independently of one another. Since only a unique set of values is possible (the nonnegative integers), the Poisson is a discrete random variable.

The Poisson distribution depends only on the average number of occurrences per unit of time or space. The mean, or average number of occurrences, for a Poisson process is λ (lambda). For example, we might have λ = 15 calls per hour coming into a telephone switchboard. The Poisson distribution has the property that the rate of occurrence can easily be adjusted to correspond to a new unit of time or space. For example, if an average of 15 calls per hour enter a switchboard, then an average of (15)·(2) = 30 calls per two hours enter the switchboard. Or an average of (15)·(1/3) = 5 calls per 1/3 hour (20 minutes) enter the switchboard. Thus, we can determine a new rate of occurrence by simple multiplication and μ_x = λt becomes the new Poisson mean.

The Poisson distribution has a general formula to generate the probability of X1 successes in time or space unit t. It is:

$$P(X1) = \frac{(\lambda t)^{X1} \cdot e^{-\lambda t}}{X1!}$$

For the switchboard example, the probability of exactly 3 calls in a 20-minute period can be determined as follows:

λ = 15 calls per hour

λt = $\frac{20}{60}$ hour = 1/3 hour

λt = 15(1/3) = 5 calls per 20-minute period

then: e = 2.71828

$$P(X1 = 3) = \frac{5^3 \cdot e^{-5}}{3!} = \frac{125 \cdot e^{-5}}{3 \cdot 2 \cdot 1}$$

$$= .1404$$

Table B in the appendix of the text provides Poisson probabilities for different values of λt. You should verify that:

$$P(X1 = 3/\lambda t = 5) = .1404$$

6-10 <u>Variance and Standard Deviation of the Poisson Distribution</u>:

The mean of the Poisson distribution has already been identified as:

$$\mu_x = \lambda t$$

In addition, <u>the Poisson has the unique distinction that the variance equals the mean</u>, so:

$$\sigma_x^2 = \lambda t$$

and therefore the standard deviation is

$$\sigma_x = \sqrt{\lambda t}$$

For the switchboard example, we have

$\mu_x = \lambda t = 15(1/3) = 5$ calls per 20 minutes

and $\quad\sigma_x^2 = 5$

and $\quad\sigma_x = \sqrt{5} = 2.236$

Note that if the mean of a Poisson random variable can be reduced, so can the standard deviation. This is a valuable advantage of the Poisson distribution, since it offers the decision maker the chance to control the uncertainty (measured by the variance or standard deviation) in his/her decision environment.

```
*****************************
*                           *
*         SOLUTIONS         *
*                           *
*****************************
```

6-1 You might have identified applications in all areas of production where it is possible to identify two, and only two, attributes of the product such as: (defective, nondefective), (smooth, rough), (clean, dirty), etc.

Other applications of the binomial occur in the areas of accounting where the accountant can identify an account balance as being in error or not or an inventory item as being present or not.

In all cases, the basic assumptions of the binomial distribution must apply. Specifically:

- n   identical independent trials
- 2   possible outcomes (success and failure)
- p   the probability of success remains constant from trial to trial

6-3 Applications which deal with a random process for which the number of successes can be counted could be listed. For instance, the distribution of hourly arrivals at hospital emergency rooms can often be considered Poisson. Other applications dealing with arrivals at service centers could be discussed.

6-5 Both distributions are discrete in nature, and it is possible to use the Poisson to approximate the Binomial distribution for large sample sizes.

6-7 P(fewer than 4) = P(3) + P(2) + P(1) + P(0). We can find these probabilities directly from the binomial table with

$$n = 20$$

$$p = .4$$

$$P(3) = .0123$$
$$P(2) = .0031$$
$$P(1) = .0005$$
$$P(0) = .0000$$
$$P(\text{fewer than } 4) = .0159$$

6-9 This distribution can be developed by using the binomial distribution tables with:

$$n = 4$$

$$p = .10$$

X1 = # of ruined diamonds	P(X1)
0	.6561
1	.2916
2	.0486
3	.0036
4	.0001
	1.0000

6-11 The variance and standard deviation for a binomial are found as follows:

$$\sigma_x = \sqrt{n \cdot p \cdot q} \quad \text{and} \quad \sigma^2 = n \cdot p \cdot q$$

$$\sigma_x = \sqrt{(4)(.1)(.9)} \quad \text{and} \quad \sigma_x^2 = (4)(.1)(.9)$$

$$\sigma_x = .6 \quad\quad\quad \sigma_x^2 = .36$$

6-13 We use the Poisson table with $\mu_x = \lambda t = 4.0$

$$P(\text{fewer than 2}) = P(1 \text{ error}) + P(0 \text{ errors})$$

$$= .0733 + .0183$$

$$= .0916$$

6-15 $E(x) = \Sigma x \, P(x)$

$$= (0)(.0498) + (1)(.1494) + (2)(.2240) \ldots (12)(.0001)$$

$$= 3.0$$

Note: the expected value of the Poisson = $\lambda t = 3.0$.

6-17 If we assume that the binomial distribution applies we can solve this problem by using the binomial table with:

$$n = 100$$

$$p = .05$$

$$P(15 \text{ or more defectives}) = P(15) + P(16) + P(17) + \ldots$$

$$= .0001 + 0000 + \ldots .000$$

Thus, if p = .05

    P(15 or more defectives) = .0001

Therefore, because this probability is so small, we conclude that the defective rate is not .05. It probably is larger than .05. In n of 100 we would expect 5 defectives but we found 15.

We will use p = .05 because that is the claimed defective rate. If the observed results have a probability which looks unreasonably low, we will conclude that the .05 defective rate is incorrect.

6-21 First, we will assume that the binomial distribution is the appropriate probability distribution, with

    n = 8 projects

    p = .15

```
                1 or 0 successes (fired)
        Buy     2 or 3 successes (okay)
                4 or more successes (bought too small)

                1 or 0 successes (fired)
        Wait    2 or 3 successes (should have bought)
                4 or more successes (okay)
```

P(1 or fewer) = P(1) + P(0)

    = .3547 + .2725 = .6572

P(2 or 3)   = P(2) + P(3)
    .2376 + .0839 = .3215

P(4 or more) = P(4) + P(5) . . . P(8)

    = .0185 + .0026 + ...0000

    = .0213

One course of action given the information presented in this problem would be to assume the outcome with the highest probability will occur. Thus, the most likely outcome is that we will get fired. Therefore, we should not buy the house now. We should wait to observe the results of the projects.

6-23 To determine the expected profits, we multiply the profit to each customer by the probability of a bolt going to that customer.

$$E(\text{Profit}) = \sum_{i=1}^{3} \text{Profit Customer i} \cdot P(\text{Customer i})$$

$$E(\text{profit}) = (1.50)(.062) + (.90)(.7853) + (.55)(.1527)$$

$$E(\text{profit}) = \$.88$$

CHAPTER 7

CONTINUOUS PROBABILITY DISTRIBUTIONS

Conceptually a continuous random variable can assume an infinity of possible values. In practice, however, we tend to round off measurements which reduces the number of possible values. Still, given any two values for a continuous variable, we can always find at least one more value between those two. This property holds because typically the values of a continuous variable are generated by a measurement process rather than a counting process as is true for a discrete random variable.

7-1 Continuous Random Variables:

Measurements of time, weight, distance, size, and volume yield truly continuous variables if the measurement device is precise enough. Many other variables which can assume a very large number of possible values such as income levels and interest rates are often modeled as continuous. In still other instances we will learn that a continuous distribution can be used to approximate a distribution which we know to be discrete.

7-2 Continuous Probability Distributions:

Continuous probability distributions are graphically represented by a smooth curve rather than a series of rectangles as occurred with discrete probability distributions. This smooth curve is appropriate since a continuous variable has so many more possible values than a discrete variable.

Continuous probability graphs must still possess two characteristics which were first introduced for discrete probability distributions: Area under the graph must still represent probability and the total area under the graph must be equal to one. To equate area and probability for continuous random variables we must restrict our attention to intervals. The probability the variable will take on a value between any two points on the continuous scale equals the area under the graph between these two points. Since area can only be defined for intervals and since an infinity of possible values exist for a continuous variable we consider the probability of some exact value occurring to be zero.

7-3 Characteristics of the Normal Distribution:

The most important continuous distribution in statistical decision making is the normal distribution. Normal distributions are unimodal with the peak occurring in the exact center of the distribution. Since normals are also <u>symmetric</u>, this insures that the <u>mean, median, and mode are all equal</u>. Normal distributions are defined for all real numbers from negative infinity to positive infinity so the graph approaches, but never really touches, the horizontal axis. This means normals are <u>asymptotic to the horizontal axis</u>.

Normal distributions are all defined by a common density function but may differ greatly in location and spread. A normal distribution becomes unique once we know the mean, μ_x, and the standard deviation, σ_x. The mean specifies the location of a normal distribution on the horizontal axis and the standard deviation indicates how the other values are spread around the mean.

7-4 <u>Finding Probabilities from a Standard Normal Distribution</u>:

All normal distributions have the same density function and differ only in location and dispersion. Therefore we can rescale any normal to the <u>standard normal distribution</u> by adjusting the location and dispersion. This convenient standardization process permits us to establish areas (probabilities) for only the standard normal and use it to compute probabilities for any normal distribution. Therefore rather than use integral calculus to determine normal curve areas (probabilities) we can use basic arithmetic and a standard normal table.

The values of a standard normal distribution are often called standard scores or normal deviates and are represented symbolically as scores. The standard normal transformation is defined by:

$$Z = \frac{x - \mu_x}{\sigma_x} = \frac{\text{value} - \text{mean}}{\text{standard deviation}}$$

This transformation produces a normal distribution which has a mean of zero and a standard deviation of one. Values of the standard normal, Z, and areas (probabilities) associated with many possible intervals are listed in Appendix C, in the text. Note the Z value is actually the number of standard deviations X is from μ_x.

The following example illustrates the use of the standard normal transformation and the standard normal table. Suppose the lifetimes of Roll Longer Tires can be described by a normal distribution with a mean of 40,000 miles and a standard deviation of 10,000 miles. If a single tire is randomly selected from all of Roll Longer's tires, what is the probability the lifetime will be at least 40,000 miles? If we let X represent the lifetime of the selected tire, then the appropriate probability can be represented by the following diagram.

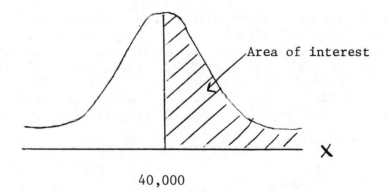

40,000

Since area corresponds to probability, we should recognize that $P(X \geq 40,000)$ = .5 because normal distributions are symmetric about the mean. Notice also that using the standard normal transformation yields

$$P(X \geq 40,000) = P(Z \geq \frac{40,000 - 40,000}{10,000}) = P(Z \geq 0),$$

which also must equal .5 since zero is the mean of the standard normal.

What is the probability that a randomly selected tire will have a lifetime between 40,000 miles and 47,500 miles? The approximate probability is represented by the area under a normal curve between 40,000 and 47,500.

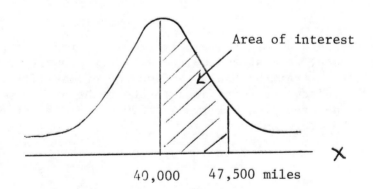

40,000   47,500 miles

We must transfer to the standard normal as follows.

$$P(40,000 < x < 47,500)$$

$$z = \frac{X - \mu_x}{\sigma_x}$$

$$z = \frac{47,500 - 40,000}{10,000}$$

$$z = \frac{7,500}{10,000}$$

$$z = .75$$

This probability can now be found directly from the standard normal table. Locate .7 in the left hand column and .05 (second decimal place) in the first row of the table. The entry at the intersection of the row containing .7 and the column headed by .05 is the probability that a tire will last between 40,000 and 47,500 miles. By inspection, this entry is .2734. Therefore P(40,000 < x < 47,500) = .2734. Since all normals are symmetric, P(32,500 < x < 40,000) = .2734. This is shown as follows:

$$Z = \frac{32,500 - 40,000}{10,000}$$

$$Z = -.75$$

From the normal table for $Z = .75$ (yields the same probability as -.75) we get .2734.

What is the probability that a randomly selected tire will last at least 55,000 miles? To solve we must again transfer to the standard normal as follows.

$$Z = \frac{55,000 - 40,000}{10,000}$$

$$Z = \frac{15,000}{10,000}$$

$$Z = 1.50$$

Now, since the standard normal table only provides areas between the value of interest and the mean, we must employ the symmetry property again to find the desired probability. From the table we find the probability corresponding to $Z = 1.50 = .4332$. Since the total area to the right of the mean = .50000, we can determine the desired probability as .5000 - .4332 = .0668. Graphically this problem can be represented as

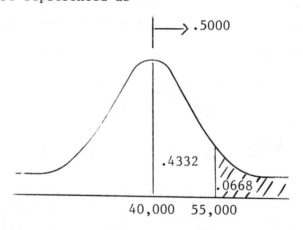

What is the probability that a randomly selected tire will have a lifetime between 37,500 miles and 45,000 miles? This problem can be represented graphically as:

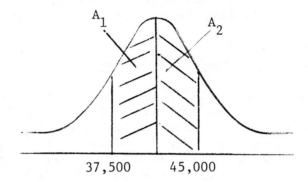

Recall that the standard normal table only provides probabilities (areas) for intervals which include the mean as an end point. Therefore we split the interval from 37,500 to 45,000 at the mean of 40,000 to yield

$$P(37{,}500 \le x \le 45{,}000) = P(37{,}500 \le x \le 40{,}000) +$$

$$P(40{,}000 \le x \le 45{,}000).$$

Now we can determine:

$$P(37{,}500 \le x \le 40{,}000) = A_1$$

$$P(40{,}000 \le x \le 45{,}000) = A_2$$

and

$$P(37{,}500 \le x \le 45{,}000) = A_1 + A_2.$$

Applying the standard normal transformation to both intervals provides:

$$z = \frac{45{,}000 - 40{,}000}{10{,}000} = \frac{50{,}000}{10{,}000} = .50$$

and

$$z = \frac{37{,}500 - 40{,}000}{10{,}000} = \frac{2{,}500}{10{,}000} = -.25$$

Then the probabilities from the normal table corresponding to these two z values are:

$$z = .50; \text{ probability} = .1915 = A_2$$

$$z = -.25; \text{ probability} = .0987 = A_1$$

To determine probabilities for normal distributions we must know the mean and standard deviation. Then it is often helpful to construct a graph which clearly identifies the areas (probabilities) of interest. Once the end points of the interval(s) of interest have been standardized, the graph helps to eliminate confusion over whether probabilities should be added to or subtracted from .5000, or possibly added to or subtracted from each other.

7-5 Other Applications of the Normal Distribution:

Sometimes it will be helpful (or necessary) to work from the standard normal distribution backwards to the original distribution. For example, a manufacturer of automobile batteries knows the lifetimes of its batteries are normally distributed with a mean lifetime of 200 weeks and a standard deviation of 20 weeks. The manufacturer wishes to establish a warranty on the batteries such that only about 10 percent of the batteries will fail before the warranty time and require replacement by the firm. As indicated by the following diagram, we wish to find a lifetime, X, such that about 10 percent of the batteries have a lifetime shorter than X. Therefore, since half (.50) of the batteries have a lifetime of 200 weeks or less, .40 of the batteries must have a lifetime greater than X but less than 200. The standard normal table provides probabilities for intervals like X to 200 which we want to specify as .40. A graph of an appropriate standard normal is

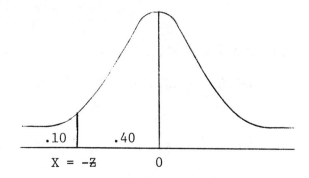

and now we must inspect the entries in the body of the standard normal table to find one as close to .4000 as possible. The entry which is closest to .4000 is .3997 which corresponds to a Z-value of 1.28. If we use the standard normal transformation, we know that $-Z = \frac{X - \mu}{\sigma_X}$ which, after substituting known values produces

$$-1.28 = \frac{X - 200}{20}$$

so $\qquad X = 200 - 1.28 (20) = 174.4$

Thus the warranty period should be about 174 weeks. Management would likely adjust the warranty to a more convenient figure like 40 months or perhaps 42 months (3 1/2 years).

When any three of the four values in the standard normal transformation of $Z = \frac{X - \mu}{\sigma_X}$ can be established, we can solve for the fourth. In this example we could establish $\mu = 200$, $\sigma = 20$, and $Z = -1.28$ so we could solve for X as 174.4. In still other cases we might be interested in μ or σ_X, but we must always have 3 of the 4 values.

7-6 Normal Approximation to the Binomial Distribution:

Even though the binomial distribution is discrete, we can often obtain a good approximation of binomial probabilities using the normal distribution. As a rule of thumb, the normal will provide an adequate approximation of binomial probabilities as long as both n·p and n·q are at least 5.

To illustrate, suppose company records indicate that 40 percent of the customers at a local store pay in cash. If a random sample of 20 transactions is selected, what is the probability that exactly 10 customers pay in cash? We should recognize this as a binomial probability of 10 successes in 20 selections when P(success) = .40. From the binomial tables, this probability is .1171. The normal approximation to this binomial probability can be found as follows:

First, $n \cdot p = (20)(.4) = 8 > 5$

$n \cdot q = (20)(.6) = 12 > 5$

so the approximation should be adequate. Now, recall that for the binomial

$$\mu_x = n \cdot p = (20)(.4) = 8$$

$$\sigma_x = \sqrt{n \cdot p \cdot q} = \sqrt{(20)(.4)(.6)} = 2.19$$

Graphically, we wish to find the area between 9.5 and 10.5 for a normal distribution with mean equal to 8 and a standard deviation of 2.19.

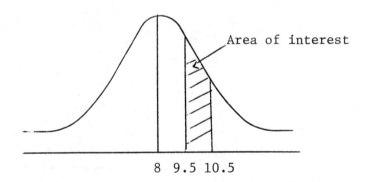

We establish the area of the interval from 9.5 to 10.5 as the probability of 10 successes because the rectangle (in a binomial probability graph) which represents the probability of 10 successes has <u>unit width</u>. The use of these one-half unit adjustments to establish an appropriate interval is referred to as the <u>continuity correction</u>. We should make a continuity correction whenever we use a continuous probability graph to approximate discrete probabilities.

Next, for a normal distribution, the $P(9.5 \leq X \leq 10.5)$ is found as follows:

$$z = \frac{9.5 - 8.0}{2.19} = .68; \text{ probability} = .2517$$

and

$$z = \frac{10.5 - 8.0}{2.19} = 1.14; \text{ probability} = .3729$$

By inspection of the standard normal table and subtraction we can then determine the approximate probability of 10 successes as .3729 - .2517 = .1212. Notice that this approximation is quite close to the true binomial probability of .1171 which was found in the binomial table.

When should we use the normal approximation to the binomial? If a binomial table for the appropriate values of n and p is available, we should use it since the probabilities are exact (subject to some rounding in the last digit). If an appropriate binomial table is not available and both np and nq exceed five, then use the normal approximation.

SOLUTIONS

7-3 (a)

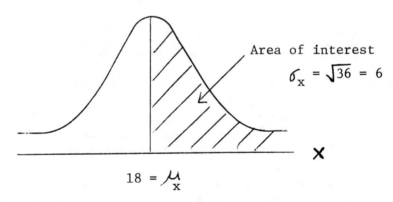

By inspection: 50 percent since the area on either side of the mean equals .50.

(b)

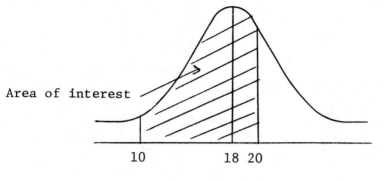

Area of interest

Area between 10 and 18

$$Z = \frac{10 - 18}{6}$$

$Z = -1.33$

Area from table = .4082

Area between 18 and 20

$$Z = \frac{20 - 18}{6}$$

$Z = .33$

Area = .1293

Total area = .4082 + .1293 = .5375

(c)

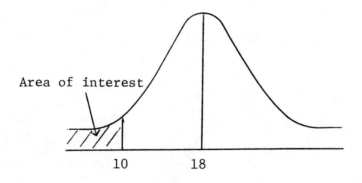

Area of interest

To find the probability of a value less than 10 we find:

.5 - area between 10 and 18

To find this area:

$$Z = \frac{10 - 18}{6}$$

$$Z = -1.33$$

Area between 10 and 18 = .4082

Area less than 10 = .5 - .4082

= .0918

7-5

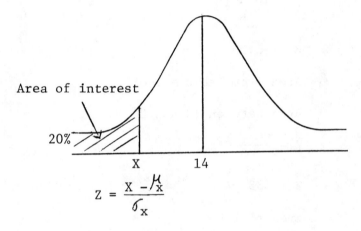

$$Z = \frac{X - \mu_x}{\sigma_x}$$

Z is still -.84, only the mean has shifted.

$$-.84 = \frac{X - 14}{6}$$

Solving for X

$$14 + (-.84)(6) = X$$

$$8.96 = X$$

7-7 (a)

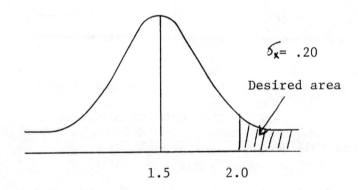

58

Prob(X > 2.0) = .5 − p(1.5 ≤ x ≤ 2.0)

$$z = \frac{2.0 - 1.5}{.20}$$

z = 2.5

From normal table:

P(1.5 ≤ x ≤ 2.0) = .4938

P(X > 2.0) = .5 − .4938 = .0062

(b)

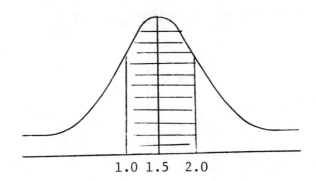

1.0 1.5   2.0

P(1.5 ≤ x ≤ 2.0) = .4938 from part (a)

Since the normal distribution is symmetrical,

P(1.0 ≤ x ≤ 1.5) also equals .4938

thus the probability:

P(1.0 < x < 2.0) = .4938 + .4938

+ .9876

(c)

P(x < 1.5) = .50 by definition.

Since the cattle gains are assumed to be independent, the probability that two cattle will both gain less than 1.5 pounds on a given day is:

(.5) · (.5) = .25

The probabilities differ in parts (a) and (b) between problems 6 and 7 due to the change in standard deviation. The smaller the standard deviation the less likely it is that an extreme value will be observed.

7-9

(a)
$$n = 300$$
$$p = .4$$
$$q = .6$$

So, using the normal approximation:

$$\mu_x = n \cdot p = (300)(.4) = 120$$
$$\sigma_x = \sqrt{n \cdot p \cdot q} = \sqrt{(300)(.4)(.6)} = 8.49$$

Remembering to adjust for integer values.

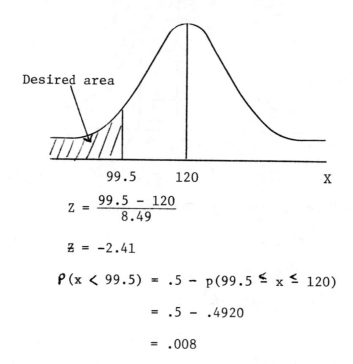

$$Z = \frac{99.5 - 120}{8.49}$$

$$Z = -2.41$$

$$P(x < 99.5) = .5 - p(99.5 \leq x \leq 120)$$
$$= .5 - .4920$$
$$= .008$$

We should analyze this problem by first finding the probability of selling 70 or fewer subscriptions given the market research results were accurate.

(b) The appropriate figure is:

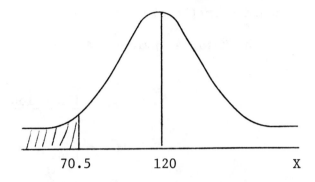

We use the normal approximation to the binomial:

$$Z = \frac{70.5 - 120}{8.49}$$

$$Z = -5.83$$

$$P(X \leq 70.5) = .5 - P(70.5 \leq x \leq 120)$$

$$= .5 - .5$$

$$= 0$$

Since this probability is essentially zero, we should conclude that either p = .4 is not an accurate estimate or the sales approach does not match that used in the market research.

(c) $E(Sales) = \mu_x = n \cdot p$

$$= 500 \cdot 4$$

$$= 2,000 \text{ sales}$$

7-11 (a) Since this is a physical production process, it is likely normally distributed. If we can assume a normal distribution we can use the material in this chapter.

(b) We know μ_x, but not σ_x. Thus our first objective is to determine σ_x assuming the distribution is normally distributed.

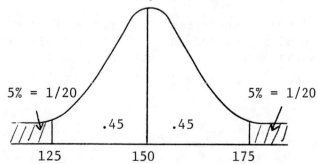

From the standard normal table:

The Z for 45 percent ≈ 1.64

then

$$1.64 = \frac{175 - 150}{\sigma_x}$$

$$\sigma_x = \frac{25}{1.64}$$

$$\sigma_x = 15.24$$

Now to compute the probability the work crew will make 100 or more feet we do the following

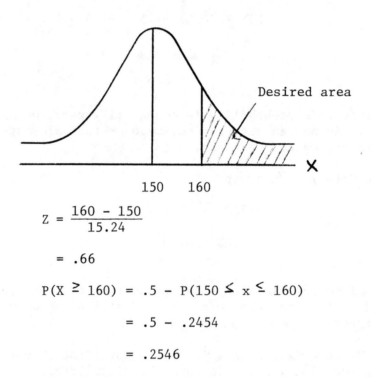

$$Z = \frac{160 - 150}{15.24}$$

$$= .66$$

$$P(X \geq 160) = .5 - P(150 \leq x \leq 160)$$

$$= .5 - .2454$$

$$= .2546$$

7-13 We must find two expected costs. First we find the expected cost of not scheduling the plumbing crew and then we compare this expected cost with the expected cost associated with scheduling the plumbers.

$$E(\text{not scheduled}) = \text{cost of needing} \times P(\text{needing})$$
$$+ (\text{cost of not needing})\,(P(\text{not needing}))$$
$$= (\$250)(.1660) = 0\,(.8340)$$
$$= \$41.50$$

$$E(\text{schedule}) = \text{cost of needing} \times P(\text{needing}) +$$
$$(\text{cost of not needing})\,(P(\text{not needing}))$$
$$= (0)(.1660) + (150)(.8340)$$
$$= \$125.10$$

Generally we would try to avoid the highest expected cost. Therefore, the best decision is to not schedule the plumbers and hope we won't need them.

7-15 We must begin by assuming the response times are independent. (This might be a heroic assumption.) Next we find the probability of a single response time exceeding 45 seconds as follows:

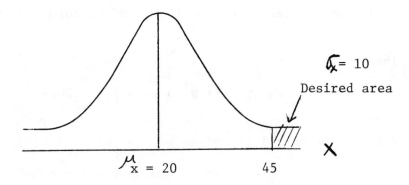

Find the number of standard deviations 45 is from the mean:
$$z = \frac{45 - 20}{10} = 2.5$$

Thus from the normal table:
$$P(X \geq 45) = .5 - .4938$$
$$= .0062$$

Then for two successive response rates of 45 or more seconds we multiply .0062 by .0062.

$$P(\text{two successive 45 second responses}) = (.0062)(.0062) = .00003844$$

This probability is extremely small which would indicate that either the manufacturer's claim is not true or that the response times were not collected under "normal" conditions.

7-17 Using μ_x = 20 seconds

σ_x = 10 seconds

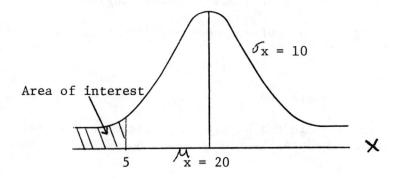

First determine the number of standard deviations 5 is from 20:

$$z = \frac{5 - 20}{10} = \frac{-15}{10} = -1.5$$

The probability associated with z = -1.5 from the normal distribution table is .4332.

Then

$$P(X \geq 5) = .5 - .4332 = .0668$$

7-19 We can use the normal approximation to the binomial as follows:

$$\mu_x = (n)(p) = (10,000)(.6) = 6,000$$

$$\sigma_x = \sqrt{(n)(p)(q)} = 48.99$$

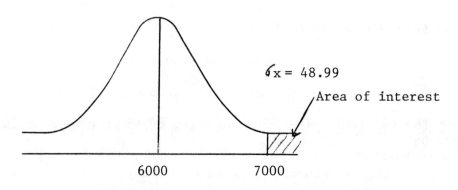

Thus

$$P(X \geq 6999.5) = ?$$

found as follows

$$z = \frac{6999.5 - 6000}{48.99}$$

$$z = 20.40$$

The probability between $z = 20.40$ and the mean is essentially .50 from the normal distribution table.

Thus

$$P(X \geq 6999.5) = .5 - .5 \approx 0$$

There is virtually no chance of observing 7000 or more businesses fail between 1980 and 1982 if the probability of failure holds at .60.

7-21 Since passenger arrivals is a yes-no situation, the true distribution is actually described by the binomial distribution. However, we can use the normal distribution to approximate the binomial but will have to adjust for discrete units. The appropriate probability is determined as follows:

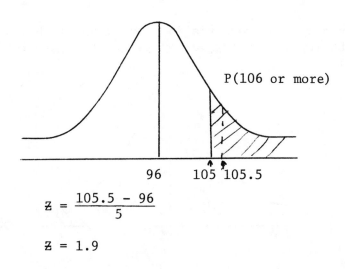

$$z = \frac{105.5 - 96}{5}$$

$$z = 1.9$$

The area for $z = 1.9$ is .4713. Therefore

$$P(106 \text{ or more}) = .5 - .4713$$

$$= .0287$$

Thus there is slightly less than a 3 percent chance of more than 105 people showing for the flight.

7-23 We will use the advertised mean and standard deviation to see if 99% of the bounces will fall between 5.85 and 6.15 inches. First:

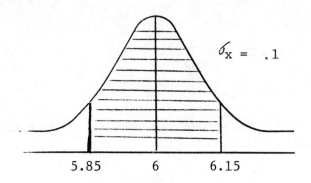

$$z = \frac{6.15 - 6.0}{.1}$$

$$z = 1.5$$

The probability of a bounce having a thickness between 6 and 6.15 inches is .4332. Since the distribution is symmetrical, the probability that a bounce will be between 5.85 and 6.0 inches is also .4332. Therefore the probability of a board being within the standards is .8664. Thus the machine, as advertised, does not meet standards.

7-25 If the mean and standard deviation are 58 and 14 hours, we are interested in the probability of finding values as great or greater than 74 and 90 respectively. First:

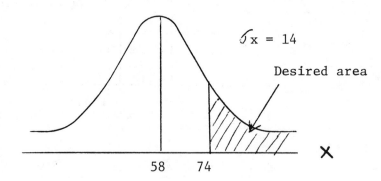

For 74 hours:

$$z = \frac{74 - 58}{14} = 1.14$$

For 90 hours:

$$z = \frac{90 - 58}{14} = 2.29$$

Since 74 is 1.14 standard deviations from the mean, we would expect a value this large or larger .1271 of the time (.5000 - .3729). However, a value of 90 is expected to occur with a probability .011 (.5 - .4890). In neither case is the probability high. Certainly, there appears to be justification for looking into the situation further. This is especially true for the person who used 90 hours of sick leave.

7-27 We will want to determine the value that is exceeded by only 10% of the workers:

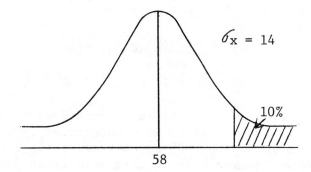

For a 10% value in the tail, $Z = 1.28$. Then we must solve for X:

$$1.28 = \frac{X - 58}{14}$$

$$X = 58 + 1.28 \times 14$$

$$= 75.92 \text{ or } 76 \text{ hours}$$

This would represent a substantial increase over the present 40 hours.

7-29 This situation is described by a binomial distribution (good and bad locks). To respond to the letter we will use the normal approximation to the binomial and find the probability of finding 74 or more defectives. First we compute:

$$\mu_x = (n)(p) = (1200)(.05) = 60$$

$$\sigma_x = \sqrt{(n)(p)(q)} = \sqrt{(1200)(.05)(.95)}$$

$$= 7.55$$

We want the probability of finding 74 or more defective locks:

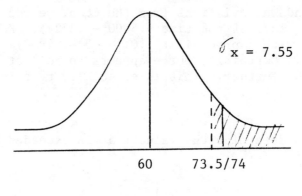

$$z = \frac{73.5 - 60}{7.55}$$

$$z = 1.79$$

The area between the mean and 1.79 standard deviations from the mean is .4633. Thus the probability of finding 74 or more defectives is .0367. There is justification for questioning the .05 percent defective rate because of this low probability. We would expect 74 or more defectives 3.67% (.5 - .4633) of the time.

7-31 We want a rental value exceeded by only 10% of the market.

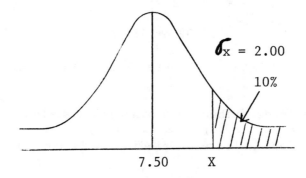

Now the Z value is 1.28. So:

$$1.28 = \frac{X - 7.50}{2}$$

$$X = 7.50 + 2(1.28)$$

$$= \cancel{\$9.06}/\text{sq. foot}$$
$$ 10.06$$

7-33 (a) This problem is solved as follows:

We want P(x ≥ 39)

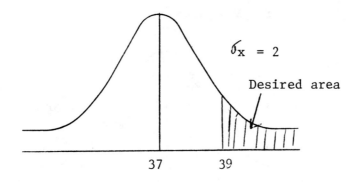

$$z = \frac{39 - 37}{2} = 1$$

Area for $z = 1$ is .3413

So:

$$P(x \geq 39) = .5 - .3413$$
$$= .1587$$

(b) We want $P(x \geq 35.5)$

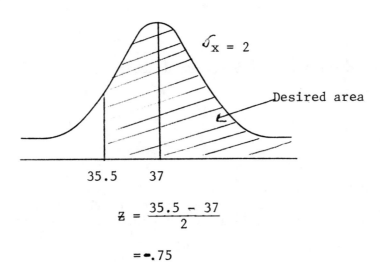

$$z = \frac{35.5 - 37}{2}$$
$$= -.75$$

Area for $z = -.75$ is .2734

So:

$$P(x \geq 35.5) = .2734 + .5000$$
$$= .7734$$

(c) We want $P(35 \leq x \leq 38.5)$

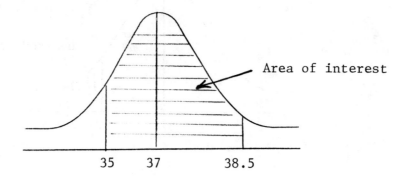

Here we need two z values:

$z_1 = \dfrac{35 - 37}{2}$ \qquad $z_2 = \dfrac{38.5 - 37}{2}$

$= -1$ \qquad\qquad $= .75$

Area = .3413 \qquad Area = .2734

To find the desired probability, add these two areas:

$P(35 \leq x \leq 38.5) = .3413 + .2734$

$= .6147$

(d) Your mileage is 6.5 standard deviations below the stated mean:

$z = \dfrac{24 - 37}{2}$

$= -6.5$

The probability of finding a value this low is essentially zero given the stated mean and standard deviation. Therefore this is evidence that the stated mean and/or standard deviation is incorrect. However, it does not prove this.

CHAPTER 8

SAMPLING TECHNIQUES

Decision makers should naturally prefer to have as much information about a population of data as is practical before making a decision. However, collecting the data can be time consuming, expensive, and sometimes the measurement process causes physical destruction of the data. For one or more of these reasons we typically sample from a population to provide information for the decision process, rather than take a census.

To collect information so that it will have the greatest value to a decision maker requires knowledge of sampling techniques. The second application of sampling techniques provides decision makers with information which can be used for statistical inferences and also provides the information at a controlled expense.

8-1 Reasons for Sampling:

Sampling, rather than taking a census, is used to gather information for several reasons. Usually the reasons for sampling can be condensed into time considerations and cost considerations. Almost always the time available before a decision must be made is less than we would prefer. Therefore, we strive to collect information so that a decision may be reached by some deadline. A time deadline often prohibits taking a census for two reasons. First, measuring every element in the population cannot be accomplished in the available time. Second, all the actual values of the population have not occurred, as is typical in forecasting or production situations, and therefore the population is inaccessible.

Cost considerations are another major reason for sampling. The measurement process can require additional people and sometimes costly equipment. The per unit cost of collecting data may be large enough to make a census economically impractical. The measurement process might also physically destroy the item under inspection as in life testing of products. If we took a census to determine the average lifetime of automobile tires we wouldn't have any tires left to sell. The expense associated with destroying all of the units in the population would be enormous.

One other reason for sampling which should be considered, is the greater potential for measurement error in a census. If we only measure a few items (sample) we can collect more detailed information and spend more time in the measurement process thereby decreasing the chances of measurement error. Since the individual collecting the data has to measure fewer items in a sample than in a census, there is also a lesser potential for errors caused by fatigue or possibly boredom.

8-2 **When to Use a Census**:

A census should not be automatically discarded as impractical or too costly. Sometimes data files are stored in a computer and all the items in the population can be examined rapidly and relatively inexpensively. Or a census may be practical if the population is not too large and measurement does not destroy the selected items.

8-3 **Fundamental Sampling Techniques**:

Sampling techniques may be classified as either __statistical__ or __non-statistical__. Statistical sampling techniques require that the elements which will be included in the sample be determined by chance or __probability__. Non-statistical sampling techniques allow for selection by __judgment__ and therefore not every element in the population has an equal chance of being selected.

__Statistical inference cannot be based on the information from a non-statistical sample__. All of the inference techniques presented in the text require some form of statistical or probability sample.

A sample is a __simple random sample__ if each element in the population has the __same chance__ of being selected. A simple random sample might be selected in a variety of ways. We might select a simple random sample by placing a slip of paper in a basket for each element in the population, mix the basket thoroughly, and then draw slips of paper to determine which elements should be included in the sample. Unfortunately, this technique really only works well for small populations.

A procedure which is similar to drawing slips from a container but can be readily used for larger populations uses a __random number table__ like Table 8-1 in the text. First we must assign a unique identification number to each element in the population. Usually this is accomplished by numbering the elements from 1 to the size of the population. Then we must randomly select a starting point in a random number table and proceed systematically through the table until the desired sample size is reached. Use as many digits from the table as there are in the largest identification numbers assigned to the population. For example, if the last number assigned to an element in the population has five digits (99999 or smaller), then select random numbers so that they contain five digits. Each random number selected from the table which is also smaller than the largest identification number for the population is associated with a unique element in the population. This element is included in the sample. If a random number repeats an earlier number or is larger than the largest identification number, discard it and select another one. Continue selecting elements until the desired sample size is reached.

Stratified random sampling, systematic sampling, and cluster sampling are alternatives to selecting a simple random sample. Each of these methods can be preferable to simple random sampling if certain conditions hold.

To use **stratified random sampling** we must be able to group elements of the population according to some characteristic. Good stratification requires that the elements within each group, or stratum, be very **similar** with respect to the measurements of interest. Different groups, or strata, should be very dissimilar.

Stratified random sampling is only more efficient (more information per dollar cost) than simple random sampling if good stratification can be accomplished. After strata are identified, a simple random sample is selected from each. In general, a larger size sample is selected from a stratum if there is a larger number of elments in the stratum, or if the variance of the elements in the stratum is large relative to other strata, or if the elements in this stratum are less costly to measure. Remember to allocate the sample size toward strata size, toward strata variability, and toward low sampling cost. Specific allocation formulas are included in many sampling texts.

Systematic sampling can be an effective alternative to simple random sampling if the elements in the population are arranged in a list. However, position in the list must be random with respect to the characteristic being measured. Alphabetic lists or a list by social security number are probably random with respect to most characteristics and thus provide a good framework for systematic sampling. To collect a systematic sample of n elements we should find k = n/N and truncate k to an integer value. Randomly select a starting point in the first k elements of the population and then select every k^{th} element after the first one. The ease with which the sample can be selected is a prime advantage of systematic sampling.

In **cluster sampling** we wish to divide the population into groups, or clusters, which are **miniature populations**. That is, we want each cluster to contain the same diversity as the population. Then we can select elements from a few randomly selected clusters and still obtain a sample which is representative of the population. For example, suppose we wish to sample the opinions of college students. A simple random sample would require that every college student have an equal chance of being selected and we might eventually have to visit hundreds of college campuses. If we can justify the assumption that all college campuses are quite similar with respect to opinion on our issue, it might be sufficient to randomly select only a few colleges and randomly sample the students on those campuses. If, as we hope, each college campus represents a miniature population, this method will provide essentially the same information as a simple random sample at a fraction of the cost.

Cluster sampling is often a viable alternative to simple random sample if natural clusters exist, but the clusters are distributed over a wide geographic area. Thus, the travel expense associated with collecting a simple random sample is prohibitive. We may require a larger sample size from each of the selected clusters but the decreased sampling cost makes a larger size sample feasible.

In contrast, stratified random sampling requires that elements within a stratum be very _similar_ and that different strata be very _dissimilar_. Then we must select items from each stratum to insure that the overall sample is representative. Cluster sampling requires that elements within a cluster be very dissimilar and that different clusters be very similar, both to each other and to the overall population. If this similarity of clusters holds, we need not sample from every group as in stratified random sampling.

Judgment sampling is an example of _non-probability sampling_. Elements are selected by the judgment of the individual collecting the sample. For individuals who are experienced at selecting judgment samples, this technique may provide very good information at a relatively low cost. However, judgment sampling does not require, nor insure, that every element in the population has some chance of being selected. Therefore no procedure exists for estimating the possible _sampling error_ and inferential statistical analysis cannot be applied.

Mail questionnaires often do not satisfy the probability conditions necessary to justify the application of statistical analysis. While we may, in fact, select the recipients of the questionnaire in a random manner, the people who actually respond often are not representative of even those who received the questionnaire, let alone representative of the population. Those who received the questionnaire but chose not to respond may (likely) hold a different set of opinions than those who did respond. We can correct for non-response bias but only if we eventually collect information from the original non-response group. If we do not call back to the original non-respondents, it is not appropriate to use statistical analysis on the sample results.

8-1 A simple random sample is one in which every possible sample of a given size has an equal chance of being selected. The selection of a simple random sample is made possible by assuring that the sampling procedure is random and the sample elements are independent. A business example in which a simple random sample would be appropriate would involve incoming product quality control. Before a batch of parts is accepted by your company, a simple random sample of parts would be selected from the batch. If this sample contained too many defectives, the entire batch could be returned to the manufacturer.

8-3 Statistical sampling approaches ranging from simple random to cluster sampling might be mentioned. Depending upon the method selected, you should be careful to consider complexities involved. For instance, how is the population defined and how will the particular sample from that population be identified? Emphasis on what is practical and what is not practical should be made.

8-5 Cluster sampling is a statistical sampling technique which usually involves dividing the population into heterogeneous groups or clusters. A random sample of these clusters is selected and then a random sample is selected from the items in each selected cluster. Business applications in which the population of interest is spread over a wide geographical area are particularly attuned to cluster sampling techniques. For example, many market research analyses begin with data collected via a cluster sampling approach.

8-7 There are several instances in which non-probability sampling might be preferred over probability sampling. For instance, there may not be enough time to survey an entire population and the decision maker is willing to survey a few key (judgmentally selected) members in hopes of learning what the other members are like. In other cases, because of the circumstances of how the population is constructed, a probability sample might not be appropriate. For example, suppose a machine which stamps out component parts in an automobile plant is responsible for hundreds of parts per hour. These parts flow down a conveyor belt and into large boxes for storage. The production people are interested in knowing whether the stamping machine has gone out of adjustment. Rather than select a random sample of parts from the storage boxes, the inspectors could simply shut down the line and inspect the latest few parts. If they are defective, the stamping machine would be adjusted or replaced.

While non-probability sampling can be very useful, the decision maker must remember that the sample results cannot be evaluated objectively using the classical techniques discussed in this text.

8-9 First of all, it is doubtful that the conclusions reached from this sample could be generalized to the larger population of skiers. Second, because of the dynamic aspects of the ski line, it might be hard to assure a random selection. Also, at any one point in time, only a fraction of the Aspen skiers are waiting in the lift line. This approach is more closely aligned to convenience sample than a statistical sample.

An attempt should be made to develop a list of the population of interest from which a random sample of names could be selected.

8-11 The clustering could be performed in stages. For instance, the first stage would be to select a random sample of states where each state is a cluster. Next, within each selected state a sample of counties could be selected and finally, phone prefixes within counties could be randomly selected for use in determining the phone numbers dialed.

8-13 Most likely, such variables as city population, city employees, or total city budget would be used to form the strata in hopes that homogeneous grouping would result and thereby reduce the variance within each stratum relative to the administrative salary variable.

Most likely, the total city budget would produce the best results assuming that administrative overhead is related to total budgets more closely than the other variables.

8-15 It is possible that the clusters could be developed such that they are not good representatives of the population as a whole. Thus instead of being heterogeneous, the clusters could contain items which are in fact homogeneous. If this occurs, it may be necessary to select more clusters and therefore a larger total sample size than if no clustering was performed.

If the elements in a cluster are homogenious, it means that these elements cannot be assumed to represent the population as a whole. Therefore more clusters will be needed to better assure that all factions of the population are represented in sample.

8-17 This answer will be entirely dependent upon your own experiences. However, you should comment on the procedures used to select the sample, define the population, and ask the questions. Relate these concepts to material discussed in this chapter and Chapter 2 of this text on sources of information.

CHAPTER 9

SAMPLING DISTRIBUTION OF $\bar{X}$

For the reasons cited in the previous chapter, sample information is a valuable aid in decision making. However, two different samples from the same population could logically support two different decisions. This apparent paradox occurs because many possible samples exist and which sample is actually selected is determined by chance. In order to develop a decision process which has a high chance of leading us to a good decision, we must recognize how many possible samples exist and learn what amount of variability can be attributed to chance sampling variation.

9-1 Relationship Between Sample Data and Population Values:

Since a <u>sample is only a subset of the population</u>, the descriptive measures which we compute from a sample will likely <u>differ</u> from the descriptive measures for the population. The sample mean, $\bar{X}$, and the population mean, μ_x, in the following example illustrate this.

An example population consists of the values 1, 2, 3, and 4. The mean of this population is $\frac{1 + 2 + 3 + 4}{4} = 2.5 = \mu_x$. If we select samples of size two without replacement from this population, we could observe six different samples ($C_2^4 = 6$). They are:

Sample	Elements	Sample Mean
1	1, 2	1.5
2	1, 3	2.0
3	1, 4	2.5
4	2, 3	2.5
5	2, 4	3.0
6	3, 4	3.5

Notice that some of the sample means equal the population mean and some do not. In general, we do not know the population mean and the sample we actually obtain is always governed by chance. Therefore <u>we anticipate that a statistic ($\bar{X}$) will not necessarily equal the corresponding parameter (μ_x)</u>.

9-2 Sampling Error:

The difference between a <u>statistic and the parameter it estimates is sampling error</u>. Since the sample we obtain is governed by chance, the size of the sampling error is also governed by chance. Therefore to characterize the sampling error we must know what values are possible for the sample mean and how likely each value is. That is, we need a probability distribution for the sample mean. The probability distribution for the earlier example is:

Sample Mean	Probability
1.5	1/6
2.0	1/6
2.5	1/6
2.5	1/6
3.0	1/6
3.5	1/6

These probabilities reflect the fact that each of the six possible samples was equally likely. A sampling distribution is a probabilitiy distribution for a statistic. Since the sample mean is a statistic, this is a sampling distribution of a mean ($\bar{X}$).

9-3 <u>Sampling Distribution of $\bar{X}$</u>:

Two properties of the sampling distribution of the mean are essential for the fundamentals of statistical inference. First, <u>the mean, or expected value, of the sampling distribution of the mean will equal the population mean</u>. The mean of all possible sample means must also equal 2.5. That is, the mean of all six possible sample means:

$$\mu_{\bar{X}} = \frac{1.5 + 2.0 + 2.5 + 2.5 + 3.0 + 3.5}{6} = 2.5$$

Thus, $\mu_{\bar{X}}$ the expected value of the sampling distribution of the mean, equals μ_x, the mean of the population from which we have sampled.

Second, <u>the standard deviation of the sampling distribution of the mean will always be less than the standard deviation of the original population</u>. We can compute the standard deviation of the sampling distribution for this example because we know all six possible sample means. The standard error of the mean is:

$$\sigma_{\bar{X}} = \sqrt{\frac{\sum_{i=1}^{k}(\bar{x}_i - \mu_{\bar{X}})^2}{K}}$$

where: K = number of possible samples

$$\sigma_{\bar{X}} = \sqrt{\frac{(1.5 - 2.5)^2 + (2.0 - 2.5)^2 + \ldots + (3.5 - 2.5)^2}{6}}$$

$$\sigma_{\bar{X}} = .645$$

Now the standard deviation of the original population is:

$$\sigma_x = \sqrt{\frac{\sum_{i=1}^{N}(x_i - \mu_x)^2}{N}}$$

$$\sigma_x = \sqrt{\frac{(1-2.5)^2 + (2-2.5)^2 + (3-2.5)^2 + (4-2.5)^2}{4}}$$

$$\sigma_x = 1.118$$

Notice, we observe less variability for the sampling distribution of the means, $\sigma_{\bar{x}} = .645$, than we do for the population, $\sigma_x = 1.118$.

9-4 **Sampling from Normal Distributions:**

In order to make probability statements about values of the sample mean we must know the shape of the sampling distribution and we must know the parameters of that distribution.

<u>If the source population possesses a normal distribution, then the sampling distribution of the mean will also possess a normal distribution.</u>

As we demonstrated in the previous section, (by example), the mean of the sample means, $\mu_{\bar{x}}$, will equal the mean of the source population. In addition, we showed that the standard error of the mean is less than the standard deviation of the population. The exact nature of this relationship is:

$$\sigma_{\bar{x}} = \sigma_x / \sqrt{n}$$

While the mean of the sampling distribution of $\bar{x}$ will always equal the mean of the source population, <u>the exact value of the standard error of the mean depends on the sample size, n.</u> If the sample size increases, the standard error of the mean decreases. We should really expect this type of relationship. If we select a large sample, we should anticipate that we will have better information about the characteristics of the population than if we select a small sample. This is exactly what the formula for the standard error of the mean indicates. <u>The standard error of the mean is a measure of expected sampling error and increasing the sample size decreases the expected sampling error.</u>

9-5 **Sampling from a Non-Normal Population:**

If the source population does not possess a normal distribution, then we must impose a minimum sample size restriction before we can be assured that the sampling distribution of the mean will be approximately normal. A commonly accepted rule of thumb for this minimum sample size is $n \geq 30$. <u>Therefore the sampling distribution of the mean will be approximately</u>

normal with $\mu_{\bar{x}} = \mu_x$ and $\sigma_{\bar{x}} = \frac{\sigma_x}{\sqrt{n}}$ regardless of the form of the population so long as the sample size is sufficiently large (n ≥ 30). This statement is generally referred to as Central Limit Theorem and is very important in the study and application of statistics.

9-6 Finite Correction Factor:

The mean of the sampling distribution of the means, $\mu_{\bar{x}}$ will equal the mean of the source population, μ_x, regardless of whether the sampling is performed with replacement or without replacement. However, if the sampling is done <u>without replacement</u>, we should make an adjustment when we compute the standard error of the mean, $\sigma_{\bar{x}}$. If we sample without replacement, then:

$$\sigma_{\bar{x}} = \frac{\sigma_x}{\sqrt{n}} \cdot \sqrt{\frac{N-n}{N-1}}$$

Notice that for our example

$$\sigma_{\bar{x}} = .645 \text{ and } \sigma_x = 1.118$$

Since sampling was performed without replacement, we can verify that

$$\sigma_{\bar{x}} = \frac{1.118}{\sqrt{2}} \sqrt{\frac{4-2}{4-1}}$$

$$\sigma_{\bar{x}} = .645$$

The term $\sqrt{\frac{N-n}{N-1}}$ is called the <u>finite correction factor</u>. The finite correction factor will always be less than one but it gets very close to one if the sample size is small relative to the population size. The closer the correction factor is to one, the less its effect on the standard error. Therefore we generally ignore the finite correction factor if the sample size is less than five percent of the population.

9-7 Decision Making and the Sampling Distribution of $\bar{X}$:

In Chapter 7 we demonstrated a procedure for finding the probability that a randomly selected value from a normally distributed population would fall between any two points. Now, armed with a knowledge of sampling and the Central Limit Theorem, we can apply the same concepts to values of the sample mean.

Press Con Manufacturing makes pre-stressed concrete beams which are used in construction. Since the quality of the raw materials and the drying process for the concrete are subject to variation, the strength of beams is a random variable. Quality control specifications require that the beams have an average strength of 50 tons with a standard deviation of 3 tons. A recent test sample of n = 36 beams yielded a sample mean strength of 48.7 tons. Does this sample evidence indicate the average strength of <u>all</u> beams might be below 50 tons? We can develop an answer to this question by determining the probability of observing a sample mean as small as 48.7 tons if the true population mean is 50 tons.

Since Press Con manufactures many beams, we will consider the sample size of 36 to be less than five percent of the population size and neglect the finite correction factor. We seek $P(\overline{X} \leq 48.7)$ if $\mu_x = 50$, $\sigma_x = 3$, and n = 36. Because of the <u>Central Limit Theorem</u> we can be confident that the sampling distribution of the mean is approximately normal since n $\geq$ 30. Therefore we can transform to the standard normal using

$$z = \frac{\overline{X} - \mu_x}{\sigma_{\overline{X}}}$$ to answer this probability question.

Thus: $$z = \frac{48.7 - 50}{3/\sqrt{36}}$$

$$z = 2.60$$

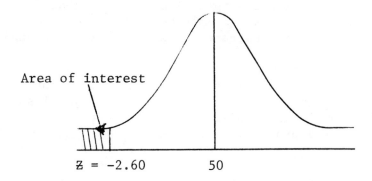

Area corresponding to $z = -2.60$ from the standard normal table is .4953. Thus

$$P(\overline{X} \leq 48.7) = .5000 - .4953$$

$$= .0047$$

This is a very small probability. Less than 5 times in 1000, in repeated sampling from a population with $\mu_x = 50$ and $\sigma_x = 3$, do we expect to observe a sample mean as small as this one. Since we only took one sample and <u>we did observe</u> $\overline{X} = 48.7$, we would infer that the population mean probably isn't as large as 50. We should monitor the production process more closely to see why the mean strength of the beams has apparently decreased.

When we ask probability questions concerning values of the sample mean, it is imperative that we consider the correct probability distribution. We must clearly distinguish between the population distribution for the individual strengths of <u>all beams</u>, the sampling distribution for the mean strengths of <u>all samples</u> of size n, and the results of <u>one particular sample</u> which we have observed.

```
*******************************
*                             *
*          SOLUTIONS          *
*                             *
*******************************
```

9-1 The Central Limit Theorem indicates that the sampling distribution of $\overline{X}$ will be approximately normally distributed with $\mu_{\overline{X}} = \mu_X$ and $\sigma_{\overline{X}} = \sigma_X/\sqrt{n}$ as long as the sample size is sufficiently large. The value of the central limit theorem is that we need not know anything about the shape of the population distribution to have an understanding about the distribution of possible sample means from which the one sample mean we obtain actually comes from. Because of this, we are able to determine the probability of the sample results using the normal distribution as the base distribution from which to compute this probability.

9-3 The reason that an increase in sample size decreases the average sampling error is that the spread in the sampling distribution of $\overline{X}$ is reduced as the sample size is increased. As the sample size is increased, it is exceedingly more difficult to obtain an extreme sample mean on either side of the population mean. Thus the chance for extreme sampling error is reduced.

9-5 If the population is normally distributed, the distribution of all possible sample means will be normally distributed also, regardless of sample size. Half of all possible sample means will lie above the true population mean and half below. Therefore the probability is .50.

9-7 The finite correction factor is used to adjust the standard error when the sampling is done without replacement and the sample size is greater than five percent of the population size.

9-9 In approaching this problem we will base our decision about what the manager should do based upon the probability of having to refund the customer's money. We determine this probability as follows:

$$\mu_X = 30$$
$$\sigma_X = 5$$
$$n = 5$$

We want: $1 - P(\overline{X} \geq 32)$

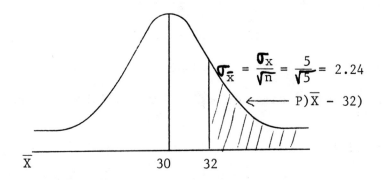

$$z = \frac{\overline{X} - x}{x} = \frac{32 - 30}{2.24} = .89$$

Then:

$$P(\overline{X} \geq 32) = .5 - .3133$$

$$= .1867$$

Thus

$$P(\text{manager must refund}) = 1 - .1867 = .8133$$

Thus with such a high probability of having to refund the money, he would recommend against the offer.

9-11 In this case, we are not discussing a random sample of n = 5. Instead we are talking about this one single customer. Thus:

$$\mu_x = 30$$
$$\sigma_x = 5$$

Now we want

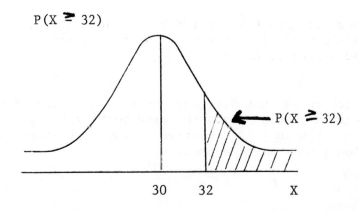

$$z = \frac{X - \mu_x}{\sigma_x} = \frac{32 - 30}{5} = .4$$

$$P(X \geq 32) = .5 - .1554$$
$$= .3446$$

The answers are different since $\sigma_{\bar{X}} < \sigma_X$. This will be true as long as $n > 1$.

9-13 For city driving:

$$\mu_X = 25$$
$$\sigma_X = 3$$

Sample Size; $n = 64$

We want:

$$P(\bar{X} \leq 21)$$

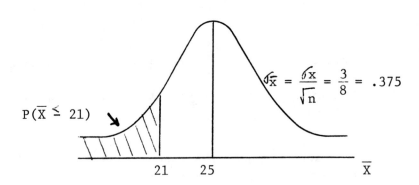

Standardizing:
$$z = \frac{\bar{X} - \mu_{\bar{X}}}{\sigma_{\bar{X}}} = \frac{21 - 25}{.375} = -10.66$$

Thus

$$P(\bar{X} \leq 21) \approx 0$$

If claimed mileage is true $P(\bar{X} \leq 21) \approx 0$. Because of this probability we should conclude the claim is not true. The true average mileage is likely to be less than the stated 25 m.p.g.

9-15 In this problem we must base our conclusion on the probability of observing the sample results given the true mean of 35 days. Note, because the sample is so large (n = 40) relative to the population size (N = 50), the finite correction factor must be used.

$$\mu_{\bar{X}} = 35$$

$$\sigma_x = 10$$
$$N = 50$$
$$\sigma_{\bar{X}} = \frac{\sigma_x}{\sqrt{n}} \sqrt{\frac{N-n}{N-1}}$$
$$= \frac{10}{\sqrt{40}} \sqrt{\frac{50-40}{50-1}}$$
$$= 1.58 \sqrt{.204}$$
$$\sigma_{\bar{X}} = .714$$

If $n = 40$ results in $\bar{X} = 46$ days, we want $P(\bar{X} \geq 46 \text{ days})$.

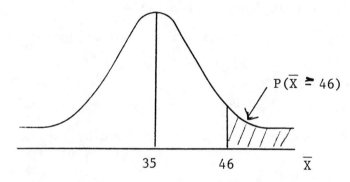

Standardizing:
$$z = \frac{\bar{X} - \mu_x}{\sigma_{\bar{x}}} = \frac{46 - 35}{.714}$$
$$= 15.4$$

Thus
$$P(\bar{X} \geq 46) \approx 0$$

If μ_x and σ_x are correct, we should conclude that customers significantly over estimate completion times.

9-17 Note, in the first part of this problem, we are concerned with an individual commercial, not a random sample of commercials. Thus:
$$\mu_x = 20,000$$
$$\sigma_x = 3,000$$

We want $P(19,500 \leq x \leq 22,000)$

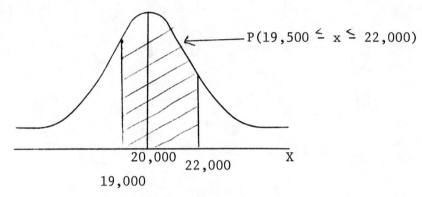

Standardizing we get:

$$z_1 = \frac{19,500 - 20,000}{3,000} = -.17$$

From the normal distribution table:

$$P(19,500 \leq x \leq 20,000) = .0675$$

$$z_2 = \frac{22,000 - 20,000}{3,000} = .67$$

$$P(20,000 \leq x \leq 22,000) = .2486$$

Thus the desired probability is found by the sum of these two probabilities:

$$P(19,500 \leq x \leq 22,000) = .0675 + .2486 = .3161$$

In the second part of this problem we want

$$P(19,500 \leq x \leq 22,000) \text{ if } n = 36$$

Note, now we are dealing with a sample mean of 36 advertisements rather than an individual advertisement. Thus:

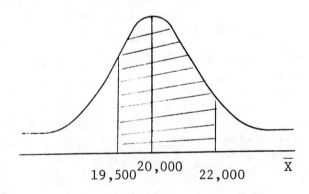

$$z = \frac{\overline{X} - \mu_x}{\sigma_{\overline{X}}}$$

where

$$\sigma_{\overline{X}} = \frac{3000}{\sqrt{36}}$$

$$= 500$$

$$z_1 = \frac{19{,}500 - 20{,}000}{500} = -1.0$$

$$P(19{,}500 \leq \overline{X} \leq 20{,}000) = .3413$$

$$z_2 = \frac{22{,}000 - 20{,}000}{500} = 4.0$$

$$P(20{,}000 \leq \overline{X} \leq 22{,}000) = .50$$

Thus

$$P(19{,}500 \leq \overline{X} \leq 22{,}000) = .3413 + .50 = .8413$$

The difference occurs because the sampling distribution for $\overline{X}$ has a much smaller spread than the distribution of the population as a whole.

9-19 We want to find the area as shown in the following figure:

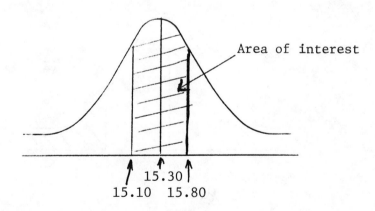

Area of interest

15.10 15.30 15.80

We will need to find two z values:

$$z_1 = \frac{15.10 - 15.30}{7/\sqrt{40}} \qquad z_2 = \frac{15.80 - 15.30}{7/\sqrt{40}}$$

$$= \frac{-.20}{1.107} \qquad\qquad = \frac{.50}{1.107}$$

$$= -.18 \qquad\qquad\qquad = .45$$

87

Area 1 = .0714   Area 2 = .1736

$$P(15.10 \leq \bar{X} \leq 15.80) = .0714 + .1736$$
$$= .2450$$

9-21 For the population:

$$\mu_x = 30.00$$
$$\sigma_x = 8.00$$

For the distribution of possible sample means:

$$\mu_{\bar{x}} = 30.00$$
$$n = 150$$
$$\sigma_{\bar{x}} = \frac{\sigma_x}{\sqrt{n}}$$
$$= \frac{8.00}{\sqrt{150}}$$
$$= .65$$

We want the area in the following figure:

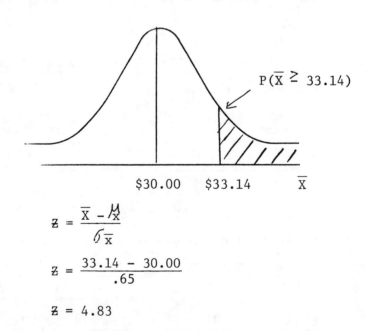

$$z = \frac{\bar{X} - \mu_{\bar{x}}}{\sigma_{\bar{x}}}$$
$$z = \frac{33.14 - 30.00}{.65}$$
$$z = 4.83$$

From the normal table area corresponding to $z = 4.83 = .5000$

Therefore

$$P(\overline{X} \geq 33.14) = .50 - .50 = 0.0 \text{ (very unlikely)}$$

If $\mu_x = 30.00$ and $\sigma_x = 5.00$, the chances of seeing an $\overline{X} = 33.14$ is essentially zero. Since we did see $\overline{X} = 33.14$, we should question the accuracy of the mean and/or standard deviation as assumed by Food King.

9-23 Here the population values are again:

$$\mu_x = 6.2$$

$$\sigma_x = 3.0$$

The sample values are:

$$\mu_{\overline{x}} = 6.2$$

$$n = 40$$

We use the finite correction factor because the sample size is large relative to the population size.

Now:

$$\sigma_{\overline{x}} = \frac{3.0}{\sqrt{40}} \sqrt{\frac{300 - 40}{300 - 1}}$$

$$= .442$$

We are interested in the area under the normal curve below 5.9 as shown in the following figure:

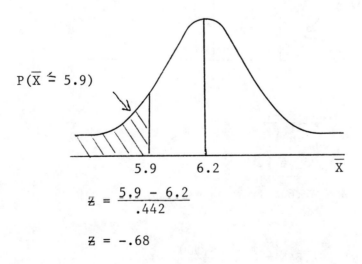

$$z = \frac{5.9 - 6.2}{.442}$$

$$z = -.68$$

Area for $z = -.68 = .2517$

So:
$$P(\overline{X} \leq 5.9) = .5000 - .2517$$
$$= .2483$$

9-25 Based upon the claim, the population values are:
$$\mu_x = 18,000$$
$$\sigma_x = 4,000$$

A sample of 64 will give a sampling distribution with the following values:
$$\mu_{\overline{x}} = 18,000$$
$$\sigma_{\overline{x}} = \frac{\sigma_x}{\sqrt{n}}$$
$$= \frac{4,000}{\sqrt{64}}$$
$$\sigma_{\overline{x}} = 500$$

We need to find the A value in the following figure which will be the cutoff point between accepting the claim and rejecting it.

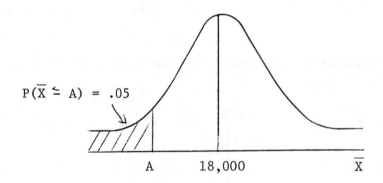

$$Z = \frac{A - \mu_x}{\sigma_x}$$

for $P(\overline{X} \leq A) = .05$ we get

$Z = -1.645$ from normal table.

So:
$$-1.645 = \frac{A - 18,000}{500}$$

Solving for A:

$$(-1.645)(500) + 18,000 = A$$

$$\cancel{\$17,277.50} = A$$

$$\$17,177.50 = A$$

9-27 (a) If the saw is set correctly, the population values are:

$$\mu_x = 120 \text{ inches}$$

$$\sigma_x^2 = .64$$

$$\sigma_x = .8 \text{ inches}$$

For a sample of 36 boards:

$$\mu_{\bar{x}} = 120$$

$$\sigma_{\bar{x}} = \frac{\sigma_x}{\sqrt{n}}$$

$$\sigma_{\bar{x}} = .8/6$$

$$\sigma_{\bar{x}} = .133$$

The area we want is shown in the figure below:

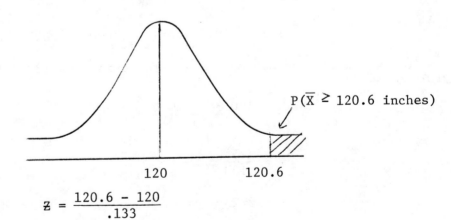

$$z = \frac{120.6 - 120}{.133}$$

$$z = \frac{.50}{.133}$$

$$z = 4.51$$

From normal table the area = .5000

So:

$$P(\overline{X} \geq 120.6 \text{ inches}) = .5000 - .5000 = 0$$

Thus there is virtually no chance of finding a mean of 36 boards which exceeds 120.6 inches.

(b) Using the same sample distribution values, we want to find the area in the following figure:

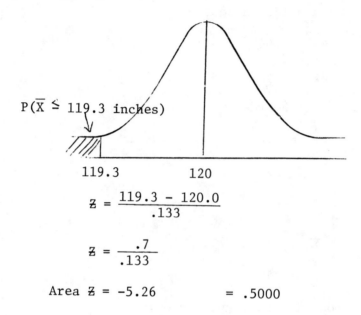

$$z = \frac{119.3 - 120.0}{.133}$$

$$z = \frac{.7}{.133}$$

Area $z = -5.26$ = .5000

So:

$$P(\overline{X} \leq 119.3 \text{ feet}) = .5000 - .5000 = 0.0$$

(c) We could find how many standard deviations 120.3 inches is from 120.0 inches:

$$z = \frac{120.3 - 120.0}{.133}$$

$$z = \frac{2.5}{.133}$$

$$z = 2.25$$

The sample mean is 2.25 standard deviations from the correct setting. The probability of observing a sample mean this far from the population mean is .0122 (very unlikely). We would probably conclude that the saw is not in adjustment.

CHAPTER 10

STATISTICAL ESTIMATION--LARGE SAMPLES

Recall from Chapter 9 that the sampling distribution of the mean is approximately normal if the sample size is greater than thirty. In this chapter we present estimation techniques which require that the sample size meets this minimum size restriction. Similar techniques are presented in Chapter 11 for the case with $n \leq 30$.

10-1 The Need for Statistical Estimation:

Managers seldom have the opportunity to take a census to provide information before making a decision. Time and cost considerations usually restrict the manager to sample information. As we observed in the previous chapter, sample information contains <u>sampling error</u>. Therefore when a manager uses sample information to aid in the decision process, errors may be committed. One of the prime advantages of the use of statistical estimation is that we can develop decision guidelines which control the chance of committing an error at some minimum level.

10-2 Point Estimation:

Point estimation is commonly used to summarize and present statistical information. <u>Simplicity of presentation is perhaps the prime advantage of point estimation</u>. Usually just a single number, or point estimate, is reported to summarize numerical data. We might read that 56% of U.S. taxpayers favor an increase in investment tax credits to stimulate business. The 56% is a <u>point estimate</u> of the true proportion of all U.S. taxpayers who favor an increase in investment tax credits and is no doubt based on sample information rather than a census. However, we infer that the population proportion is very close to 56%.

From a statistical perspective we should be concerned about insuring that the estimation procedure we choose produces "good" point estimates. Three criteria which can be used to select good point estimators are <u>unbiasedness, consistency, and efficiency</u>.

<u>A statistic is an unbiased estimator if, on the average, it equals the parameter of interest</u>. The sample mean is an unbiased estimator of the population mean since the mean of all sample means, $\mu_{\bar{x}}$, equals the mean of the source population, μ_x.

A statistic is a consistent estimator if the expected sampling error decreases as the sample size increases. Recall that the standard error of the mean, $\sigma_x/\sqrt{n}$, is a measure of sampling error. Since the sample size, n, appears in the denominator, the standard error decreases as n increases. Therefore the sample mean is a consistent estimator of the population mean.

A statistic is an efficient estimator if its sampling distribution has a "small" standard error. Usually "small" is determined by comparison with another estimator. That is, if two estimators of a parameter are unbiased and consistent, we should prefer the one whose sampling distribution has a smaller standard error. For example, if the population is symmetric, both the sample mean and the sample median are unbiased and consistent estimators of the population mean. However, the sampling distribution of the mean exhibits less variability (has a smaller standard error) than the sampling distribution of the median. Therefore the sample mean is a more efficient estimator of the population mean than the sample median.

The sample mean was used in this discussion simply as a focal point. Many other statistics also possess the necessary qualities to make them "good" estimators. A disadvantage of point estimation is that we usually do not communicate how "good" we feel the estimate is. A procedure which helps communicate the quality of the inference is interval estimation.

10-3 Confidence Interval Estimation:

The term interval estimation is quite descriptive since it really means that we will construct an interval and we estimate the parameter as any value in between the end points of the interval. The width of the interval is a function of expected sampling error and therefore helps to communicate the quality of the inference. Another factor besides sampling error which affects the width of the interval is a measure of our faith in the interval estimation procedure. This faith is expressed in the form of a confidence level.

The general format for all the confidence intervals presented in this chapter is:

$$\text{point estimate} \pm \left(\begin{matrix}\text{interval}\\\text{coefficient}\end{matrix}\right)\left(\begin{matrix}\text{standard}\\\text{error}\end{matrix}\right)$$

In this expression we multiply an interval coefficient by the appropriate standard error to provide a measure of expected sampling error. The interval coefficient is selected to establish a confidence level and the standard error is usually computed from the sample information.

10-4 Confidence Interval Estimation of μ_x--Large Samples, σ_x Known:

The general format for a confidence interval can be tailored to a specific formula for estimating the expected value of nearly any population. With σ_x known and a sample size greater than thirty, the population mean is estimated using

$$\overline{X} \pm z \cdot \sigma_{\overline{X}}$$

where $\overline{X}$ is a point estimate, the sample mean

$\sigma_{\overline{X}} = \dfrac{\sigma_x}{\sqrt{n}}$ is the standard error of the mean

and z is a standard normal score chosen to establish a specified confidence level.

Thus specific formula also assumes the sample size is less than five percent of the population size so that the finite correction factor is not required.

The standard normal score can be selected to yield any confidence level between 0% and 100%. For most applications the desired confidence level falls in the range of 80% to 99%.

In general, we wish to establish an interval which we are very confident contains the parameter of interest and yet have only a small set of possible parameter values. That is, <u>we desire a very precise, or very narrow, confidence interval</u>.

If we examine the specific confidence interval formula for estimating a population mean, we can see that two factors influence how precise, or narrow, our interval will be. Since the width of any interval is determined by the product $z \cdot \dfrac{\sigma_x}{\sqrt{n}}$, <u>only the standard interval coefficient, z, and the sample size, n, can be adjusted to change the width of the interval</u> (σ_x is a constant).

If we want greater confidence that our interval is correct (contains the parameter of interest), we select a larger interval coefficient from the standard normal table and construct a wider interval. Thus <u>increasing our confidence level makes the interval wider and less precise</u>.

If we want increased precision (a narrower interval) at a given confidence level, we select a larger size sample. Since the sample size is in the denominator of the formula for the standard error, <u>increasing the sample size makes the product</u> $z \cdot \dfrac{\sigma_x}{\sqrt{n}}$ <u>smaller and therefore the resulting interval is narrower</u>.

So our two objectives, high confidence and narrow (precise) intervals, are in conflict with one another. Fortunately, increasing the sample size allows us to maintain a specified confidence level and still obtain a specified level of precision. Unfortunately, sampling may be costly and we may have to sacrifice precision and confidence to control cost.

A formula for determining an adequate sample size to estimate μ_x with specified confidence and precision is $n = \dfrac{z^2 \sigma_x^2}{e^2}$.

In this formula σ_x is either assumed known or is the best available estimate of the population standard deviation and e is the maximum tolerable sampling error. Notice that both confidence level and precision can be controlled since Z and e are both represented in the sample size formula.

10-5 Confidence Interval Estimation of μ_x — Large Samples, σ_x Unknown:

Very seldom do we know the population standard deviation, σ_x. When σ_x is unknown we substitute the sample standard deviation in the confidence interval formula, yielding $\overline{X} \pm Z \cdot S_x/\sqrt{n}$. As long as the sample size is relatively large the sample standard deviation will provide a reasonable estimate of σ_x and the interval estimation procedure will still be valid.

For example, if we tested 100 steel belted radial tires from a production run of 100,000 and found a mean lifetime of 42,500 miles with a standard deviation of 10,000 miles, a ninety percent confidence interval for the mean lifetime of all 100,000 tires would be:

$$42,500 \pm 1.645 \left(\frac{10,000}{\sqrt{100}}\right)$$

$$42,500 \pm 1,645$$

or

$$40,855 \text{ ——— } 44,145$$

We are fairly confident that the true mean lifetime falls in this interval because we know that approximately 90 percent of the intervals we might construct in this manner would enclose μ_x.

If we desire greater confidence that our interval includes μ_x, we could select a larger interval coefficient (i.e. Z value). We must realize, though, that the resulting interval would be less precise (wider). To maintain our precision and increase confidence, we must select a larger sample.

10-6 Confidence Interval Estimation of a Population Proportion:

We can rely on the Central Limit Theorem to develop an estimation procedure for p, the proportion of successes in a binomial population. As in the previous section our general formula of point estimate $\pm$ (interval coefficient) (standard error) can be tailored for this specific estimation problem.

The point estimate of a population proportion is the proportion of successes we observe in a sample. Thus

$$\hat{p} = \frac{x}{n} = \frac{\text{number of successes}}{\text{sample size}}$$

is used to provide a point estimate of p. The interval coefficient is again a standard normal score chosen to reflect a specified level of confidence. The standard error must be estimated using sample data and is defined as

$$S_{\hat{p}} = \sqrt{\frac{\hat{p}\hat{q}}{n}}$$

where $\hat{q} = 1 - \hat{p}$, or $\hat{q}$ is the proportion of failures in the sample. The confidence interval formula becomes:

$$\hat{p} \pm z \cdot \sqrt{\frac{\hat{p}\hat{q}}{n}}$$

To illustrate, suppose a county has mandatory automobile inspection when cars are licensed and we are confident that the cars inspected on any one day approximate a random sample of all cars in the county. If 20 of the 100 cars inspected on a selected day have inadequate tire pressures, then we could establish a 95% confidence interval for the proportion of all registered cars with inadequate tire pressure as

$$.20 \pm 1.96 \sqrt{\frac{(.20)(.80)}{100}}$$

or

$$.20 \pm .0784$$

and we estimate, with 95% confidence, that between .1216 and .2784 of the registered cars in our county are operating with inadequate tire pressures.

If we wish to specify the precision and confidence level of our estimate before sampling, we can determine the necessary sample size using

$$N = \frac{z^2 pq}{e^2}$$

where p is the best available estimate of the population proportion and $q = 1 - p$. Just as when we determined the necessary sample size to estimate μ_x, z is the interval coefficient from the standard normal distribution and e is the tolerable sampling error. If we have no available value for p, we should use $p = .5$. Using $p = .5$ will provide a conservatively large sample size since pq assumes its largest value of .25 at that point. Any other value of p makes pq smaller than .25 and the resulting sample size will also be smaller.

10-7 <u>Confidence Intervals for Estimating the Difference Between Two Population Parameters--Large Samples</u>:

Is the mean tire lifetime greater for Firestone Tires or for Goodyear Tires? Does a greater proportion of the registered voters in California

97

favor federal income tax cuts than their counterparts in New York? While these questions require inferences about means and proportions, respectively, they also require a comparison of two populations. While the general confidence interval formula of (point estimate) $\pm$ (interval coefficient)(standard error) still provides the basis for inference, both the point estimate and the standard error must combine the information from two samples (populations).

Difference Between Two Population Means:

The specific formula to construct a confidence interval for the difference between two population means is

$$(\bar{X}_1 - \bar{X}_2) \pm z \sqrt{\frac{s_1^2}{n_1} + \frac{s_2^2}{n_2}}$$

Notice that even though we are estimating the difference in two means, the variances are summed under the square root in determining the appropriate standard error.

To illustrate, a national retailer collected random samples of male and female credit card holders to allow a comparison of their mean account balances. The 64 male cardholders had a mean balance of $125 with a standard deviation of $40. The 100 female cardholders had a mean balance of $110 with a standard deviation of $30. A 95% confidence interval for the difference in mean credit card balance is then

$$(\$125 - \$110) \pm 1.96 \sqrt{\frac{40^2}{64} + \frac{30^2}{100}}$$

$$15 \pm 1.96 \, (5.83)$$

$$15 \pm 11.43$$

$$3.57 \qquad \qquad 26.43.$$

Therefore we estimate, with 95% confidence, that male credit cardholders have a mean balance from $3.57 to $26.43 greater than female cardholders.

Difference Between Two Population Proportions:

The general confidence interval formula can also be tailored to estimate the difference between two proportions. This specific formula is

$$(\hat{p}_1 - \hat{p}_2) \pm z \sqrt{\frac{\hat{p}_1 \hat{q}_1}{n_1} + \frac{\hat{p}_2 \hat{q}_2}{n_2}}$$

Notice that the terms under the square root are summed even though we take the difference between the sample proportions to develop the point estimate.

The proportion of athletes who eventually obtain a degree is a subject of concern for most colleges and universities. A random sample of 100 athletes who had competed in the East Conference contained 40 who obtained a degree. A random sample of 100 athletes who had competed in the West Conference contained 50 who obtained a degree. To determine whether the proportion of all athletes who graduate may differ for these two conferences, we can construct a confidence interval for the difference between proportions. At the 90% confidence level this interval is

$$(.4 - .5) \pm 1.645 \sqrt{\frac{(.4)(.6)}{100} + \frac{(.5)(.5)}{100}}$$

$$-.10 \pm .115$$

$$-.215 \text{ ——————— } .015.$$

Since this interval includes zero, it is possible that the proportion of athletes receiving degrees in the two conferences could be the same (if $p_1 - p_2 = 0$, then $p_1 = p_2$).

10-8 <u>Conclusions</u>:

Interval estimation for a single mean, a single proportion, the difference between two means, or the difference between two proportions always starts with the general formula of

(point estimate) $\pm$ (interval coefficient)(standard error).

<u>In general, increasing the confidence level makes the interval less precise (wider) and increasing the sample size(s) makes the interval more precise (narrower)</u>.

```
****************************
*                          *
*         SOLUTIONS        *
*                          *
****************************
```

10-1 A point estimate is a single measure computed from a sample. The point estimate provides a point estimate of the associated population parameter. An <u>interval estimate</u> is a range with limits above and below the point estimate between which the population parameter does or does not fall. For each possible sample that could be selected, a confidence interval estimate could be developed. Of these intervals, a percentage of them equal to the confidence level will include the parameter.

10-3 You should recall that we have no reason to believe that a point estimate will exactly equal the parameter due to sampling error. Consequently, in many business applications, it is desirable to report an interval estimate for which a confidence level is attached with respect to the likelihood that it includes the parameter.

10-5 The equation is

$$\bar{X} \pm Z \frac{\sigma_X}{\sqrt{n}}$$

where

$\bar{X}$ = point estimate

Z = interval coefficient

$\dfrac{\sigma_X}{\sqrt{n}}$ = standard error

10-7 Precision, as defined in this chapter, is the total width of the confidence interval. Precision is the difference between the upper and lower limit of the confidence interval.

10-9 Since precision is defined to be the total width of the confidence interval, it is affected by the level of confidence chosen. For instance, if the confidence level is increased from 90% to 95%, the interval coefficient (Z value) must increase. If Z increases, the total width of the interval increases and thus precision is decreased.

10-11 We would disagree with this statement. The parameter either falls in the interval or it does not. A parameter is not a random variable, rather it has a fixed value. However, of all possible intervals, 95% will include the population value providing the interval coefficient corresponds to the 95% level.

10-13 The general format for a confidence interval is:

$$\bar{X} \pm Z\, S_x/\sqrt{n}$$

The following statistics are determined:

$$\bar{X} = 14{,}205$$

$$S_{\bar{x}} = \frac{S_x}{\sqrt{n}} = \frac{1010}{\sqrt{64}} = 126.25$$

$$Z_{.90} = 1.645 \text{ from normal table}$$

Thus:

$$14{,}205 \pm 1.645\ (126.25)$$

$$14{,}205 \pm 207.68$$

$$13{,}997.31 \text{———} 14{,}412.81.$$

The CHP should be 95% confident that the true average number of vehicles passing the off ramp is between 13,997 and 14,412 per day.

10-15 The CHP can be 99% confident that this interval includes the population mean. It is noted that the increase in confidence level results in decreased precision (wider confidence interval).

Thus the Desired Precision = 300

Tolerable Error = 150

where

$$e = \text{tolerable error}$$

$$e = z\ S/\sqrt{n}$$

then

$$150 = (2.575)\ \frac{1010}{\sqrt{n}}$$

Solving for n, we get

$$n = \frac{z^2 s^2}{e^2}$$

Then

$$n = \frac{(2.575)^2\ (1010)^2}{(150)^2} = 300.6$$

$$n = 301$$

Thus the required sample size assuming S remains equal to 1010 and 95% confidence and precision equal to 300 is 301 days.

10-17 The sample which was the largest is sample 3. This can be determined by examining the width of the confidence interval since sample size affects the standard error and the standard error impacts on the width of the interval.

Specifically, holding everything else constant, the larger the sample size, the narrower the confidence interval.

For the same reason, the smallest sample is sample 2 because it has produced the widest interval.

$$\text{largest sample} = 3 \text{ (greatest precision)}$$

$$\text{smallest sample} = 2 \text{ (least precision)}$$

10-19 $n = 60$

$z = 1.28$

sample 3 = smallest standard deviation (it has the greatest precision)

10-21 Since the means fall at the center point of the confidence interval, we determine the means as follows:

Sample 1 $\bar{X} = 39 + \dfrac{50 - 39}{2} = 44.5$

Sample 2 $\bar{X} = 40 + \dfrac{51.9 - 40}{2} = 45.95$

Sample 3 $\bar{X} = 42 + \dfrac{50 - 42}{2} = 46$

Therefore Sample 1 has the smallest sample mean.

10-23 $n = 120$

$\bar{X} = 17.56$

$S_X = 4.5$

(a) $\bar{X} \pm z \dfrac{S_X}{\sqrt{n}}$

for 94% confidence $z = 1.88$

$17.56 \pm 1.88 \dfrac{4.5}{\sqrt{120}}$

$17.56 \pm .77$

$16.79 \text{ --------- } 18.33$

(b) While good business practice would no doubt, dictate that the manager placate the customer, he should take the claim lightly since there is virtually no chance of such a long wait occurring given the sample information. This is determined as follows: Assume the higher limit for the true mean (i.e. $\mu_x = 18.33$ minutes), then if we also assume a normal distribution:

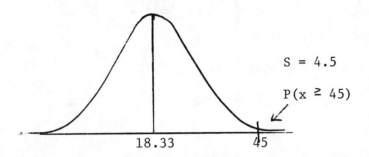

$P(x \geq 45)$ is found as follows:

$$z = \frac{45 - 18.33}{4.5} = 5.92$$

Area to the right of $z = 5.92$ is very small or practically 0.0.

(c) Such factors as time of day the sample was taken; the day of the week; season of the year. Also, was the sample random?

10-25 (a) $n = 350$

for a 97% confidence level, $z = 2.17$

$$\hat{p} = \frac{89}{350} = 0.254$$

The desired confidence interval is:

$$\hat{p} \pm z\, \sigma\hat{p}$$

$$0.254 \pm 2.17 \sqrt{\frac{0.254 \times 0.746}{350}}$$

$$0.254 \pm .052$$

.228 to .332

Thus the manager is confident that the actual proportion of clear redwood is between .228 and .332.

(b) If he finds 249 standard, there were 101 clear boards. So of the total 700 boards, 190 were clear. The desired confidence interval is

now
$$\hat{p} \pm z\sigma_{\hat{p}}$$

$$\hat{p} = \frac{190}{700} = .271$$

$$\sigma_{\hat{p}} = \sqrt{\frac{.271 \times .729}{700}} = .017$$

The interval is

$$.271 \pm 2.17 \, (.017)$$

$$.271 \pm .036$$

$$.234 \text{ to } .308$$

Thus the precision has been improved since this interval is narrower than that determined in part (a).

10-27 A response to the district manager's claim might be phrased:

The sample size of 180 is indeed adequate to draw a meaningful conclusion. In fact, a sample size smaller than 180 could also provide meaningful conclusions. The definition of what is meaningful centers on the desired precision of the estimate. A larger sample size will improve precision (decrease interval width). Therefore, if the district manager is unhappy about the resulting precision of the estimate, he can specify the precision level he desires and the required sample size can be determined. For example:

For precision of .04 (tolerable error = e = .02), we want $\hat{p} \pm .02$.

Therefore $.02 = z\sigma_{\hat{p}}$

Then if we use $\hat{p} = .65$ and a 96% confidence level, we solve for n as follows:

$$.02 = 2.05 \sqrt{\frac{.65 \times .35}{n}}$$

$$n = \frac{z^2 \hat{p} \hat{q}}{e^2}$$

$$n = \frac{(2.05)^2 (.65)(.35)}{(.02)^2}$$

$$n = 2{,}390$$

Thus, for $\hat{p} = .65$ and a 96% confidence level, the required sample size to generate a precision level of .04 is 2,390 automobiles.

10-29 The precision is .15 (i.e. e = .075), the confidence interval must be

$$\bar{X} \pm .075$$

therefore

$$.075 = z \frac{\sigma_x}{\sqrt{n}}$$

if

$$z = 1.645,$$

$$\sigma_x = 1.52$$

Solving for the required sample size we get

$$.075 = 1.645 \frac{1.52}{\sqrt{n}}$$

$$n = \frac{z^2 \sigma_x^2}{e^2}$$

$$n = \frac{(1.645)^2 (1.52)^2}{(.075)^2}$$

$$n = 1111$$

Thus, the required sample size is 1111.

This interval only refers to average protection. Some people will be protected for longer and some for shorter periods of time.

10-31 To help make this decision, we will develop a 95% confidence level. (You might have selected a different confidence level.) Based upon this confidence level, a 95% confidence interval will be developed to estimate the difference between the average hardness coefficients for the two types of materials. If this interval includes 0 it would be reasonable for manufacturing departments to select Type II over Type I if that is their preference.

The interval formula is:

$$(\bar{X}_1 - \bar{X}_2) \pm z \sqrt{\frac{S_1^2}{n_1} + \frac{S_2^2}{n_2}}$$

$$\bar{X}_1 = .66 \qquad \bar{X}_2 = .64$$

$$S_1 = .08 \quad S_2 = .063$$
$$n_1 = 64 \quad n_2 = 64$$

For 95% confidence, $z = 1.96$.

The desired interval is:

$$(.66 - .64) \pm 1.96 \sqrt{\frac{.08^2}{64} + \frac{.063^2}{64}}$$

$$.02 \pm .0249$$

$$-.00494 \text{ ——————— } .04494$$

The confidence interval contains 0. Therefore we cannot conclude a difference exists. Thus Type II can be selected, based upon this data, if desired.

10-33   (a) The approach will be to develop a 95% confidence interval estimate of the percent of new blenders that will be returned under the one year warranty. If this confidence interval includes 3%, we would conclude that the claim about the new switch is correct.

The confidence interval formula is:

$$\hat{p} \pm z \, \sigma_{\hat{p}}$$

$$\hat{p} = \frac{9}{250} = .036$$

$$\sigma_{\hat{p}} = \sqrt{\frac{.036 \times .964}{250}} = .012$$

For 95% confidence level, $z = 1.96$

The desired interval is:

$$.036 \pm 1.96 \, (.012)$$

$$.036 \pm .024$$

$$.012 \text{ ——————— } .069$$

Since the 95% interval of .012 to .06 contains 3%, we cannot reject the claim about the switch mechanism.

(b) We can combine the sample information for the new blenders with this sample information and develop a confidence interval estimate for the difference between proportions returned for the new versus old blenders.

We use:

$$(\hat{p}_1 - \hat{p}_2) \pm z \sqrt{\frac{\hat{p}_1 \hat{q}_1}{n_1} + \frac{\hat{p}_2 \hat{q}_2}{n_2}}$$

where: new   old

$$\hat{p}_1 = \frac{9}{250} = .036 \qquad \hat{p}_2 = \frac{16}{250} = .064$$

$$n_1 = 250 \qquad n_2 = 250$$

The desired interval is:

$$(.036 - .064) \pm 1.96 \sqrt{\frac{.036 \times .946}{250} + \frac{.064 \times .936}{250}}$$

$$(-.028) \pm .038$$

$$-.066 \text{———} .010$$

Since the 95% interval for the difference between the two population proportions contains 0, we cannot conclude that a difference exists between the switch types, based on this sample data.

10-35 (a) We are estimating a population proportion. The appropriate interval is:

$$\hat{p} \pm z \, \hat{\sigma}_{\hat{p}}$$

where

$$\hat{p} = \frac{38}{50} = .76$$

and

$$\hat{\sigma}_{\hat{p}} = \sqrt{\frac{\hat{p}\hat{q}}{n}}$$

$$= \sqrt{\frac{.76 \times .24}{50}}$$

$$= .06$$

Assuming a 98% confidence level, the interval coefficient is $z = 2.33$.

The interval estimate is:

$$.76 \pm 2.33 \, (.06)$$

$$.76 \pm .1398$$

$$.6202 \text{———} .8998$$

(b) Since the lower limit of the confidence interval is greater than .50, we can conclude the majority prefer Chrysler with 98% confidence.

10-37 We wish to estimate the difference between two population means. The appropriate confidence interval formula is:

$$(\bar{X}_1 - \bar{X}_2) \pm z \sqrt{\frac{S_1^2}{n_1} + \frac{S_2^2}{n_2}}$$

The z value for a 94% confidence interval is 1.88.

$$\bar{X}_1 = 42{,}156$$

$$\bar{X}_2 = 43{,}414$$

$$S_1 = 3{,}455$$

$$S_2 = 2{,}981$$

The desired interval is:

$$(42{,}156 - 43{,}414) \pm 1.88 \sqrt{\frac{3455^2}{64} + \frac{2981^2}{55}}$$

$$-1258 \pm 1.88 \,(589.98)$$

$$-1258 \pm 1109.17$$

$$-2367.17 \longrightarrow 148.82$$

Since the interval does not contain 0, at the 94% confidence level, we should prefer Brand M as this brand would appear to produce longer wearing tires on the average.

CHAPTER 11

STATISTICAL ESTIMATION--SMALL SAMPLES

If a decision maker is unable to comply with the minimum sample size requirements presented in the preceding chapter, then other inference techniques must be employed.

11-1 Small Sample Estimation:

In Chapter 10 we learned to construct interval estimates for a single population mean using

$$\overline{X} \pm z \left(\frac{S_x}{\sqrt{n}}\right).$$

We recognize that both $\overline{X}$ and S_x are sample values and therefore are subject to sampling variation. However, since we restricted our attention to large samples, we argued that S_x should be very close to σ_x and sampling variation for S_x could be discounted. We also relied heavily on the <u>Central Limit Theorem</u> to insure that the sampling distribution of the mean was nearly normal thus allowing us to use standard normal scores for our interval coefficients regardless of the form of the population.

If the sample size is small, we must question whether sampling variations for S_x can be ignored and whether standard normal interval coefficients are still appropriate. The answer is <u>no</u> in both instances. Fortunately, a small sample procedure, similar to the large sample procedures of Chapter 10, is available which addresses both of these issues.

While a unique distinction between "large sample size" and "small sample size" does not exist, a common rule of thumb does exist. <u>Usually $n \geq 30$ constitutes a "large" sample and n < 30 is used to identify a "small" sample.</u>

The general confidence interval formula of:
(point estimate) $\pm$ (interval coefficient)(standard error)
is still appropriate for small sample inference. And we will show that, at least computationally, <u>the only difference between large sample confidence intervals and small sample confidence intervals is the procedure for determining the interval coefficient.</u>

11-2 The Student t Distribution:

A distribution which takes into account sampling variation for both the mean and the standard deviation is the Student t. However, the Student t distribution is only appropriate if the sampled population is normally distributed. Therefore all the procedures in this chapter require that we be able to justify the assumption that the population(s) of interest is(are) normally distributed.

More accurately, the Student t is really a family of distributions. While all t distributions are symmetric about a mean of zero, the dispersion of a t distribution depends on its degrees of freedom. For the single sample procedures discussed in this chapter, degrees of freedom is defined as sample size minus one (n - 1). As the degrees of freedom for a t distribution increase, the variance of the distribution decreases. Intuitively this decrease in dispersion occurs because as the sample size (degrees of freedom) increases, we observe less sampling variation.

Fortunately the behavior of the family of t distributions is well known and standardized tables are presented in the appendix of the text. Two values are needed to assess the proper student t values from the t table. First we must know the degrees of freedom to identify the appropriate row in the table. Second we must locate the appropriate column for confidence intervals by using the desired confidence level.

Notice that as we progress down any column in the table, the degrees of freedom increase and the tabled values decrease. This reflects the decreased variance (dispersion) for larger sample sizes. In fact, as the degrees of freedom increase, the tabled values approach the standard normal score which would be used as an interval coefficient at each of these confidence levels. This is why we can use standard normal scores for large sample confidence intervals even though σ_x is unknown.

11-3 Estimating the Population Mean--Small Samples:

The specific formula to construct a small sample confidence interval for a single mean is

$$\overline{X} \pm t \left(\frac{S_x}{\sqrt{n}}\right).$$

You should recognize that only the interval coefficient has changed from the formula used in the previous chapter. In addition, we must be certain to investigate whether the form of the population is normal since that is a requirement underlying this procedure.

Prestressed concrete beams are designed to support great amounts of weight. The beams must be broken to determine just how much weight they will support. Destructive sampling of this type almost always severely limits the size of the sample.

Suppose we test (break) five beams and determine the mean breaking weight to be 160 tons with a standard deviation of 15 tons. If we can assume that the distribution of breaking weights for all beams produced by this process is normal, then a 90% confidence interval for the population mean is

$$160 \pm 2.132 \left(\frac{15}{\sqrt{5}}\right)$$

or

$$160 \pm 14.3$$

$$145.7 \longrightarrow 174.3$$

We estimate, with 90% confidence, that the true mean breaking strength for all beams of this type falls between 145.7 tons and 174.3 tons.

11-4 Estimating the Difference Between Two Population Means--Small Samples:

The procedure for inferences about two population means using small samples requires that (1) <u>the two samples be randomly and independently selected</u>, (2) <u>both populations are normally distributed</u>, and (3) <u>the two populations have equal standard deviations</u>.

We can insure that we will have random and independent samples by the manner in which we collect the data. The last two conditions may require statistical validation. Techniques are presented in subsequent chapters of the text which could be used to check for normality and equality of standard deviations.

If these three conditions are met, the specific formula for estimating the difference in population means is:

$$(\overline{X}_1 - \overline{X}_2) \pm t \cdot S_p \sqrt{\frac{1}{n_1} + \frac{1}{n_2}}.$$

While the general format is similar to the procedure for large samples, we must now compute Sp, <u>a pooled estimate of the common standard deviation</u>. Since, by assumption, the population standard deviations are the same, we combine the information from the two samples using

$$S_p = \sqrt{\frac{S_1^2 (n_1 - 1) + S_2^2 (N_2 - 1)}{n_1 + n_2 - 2}}$$

The interval coefficient in this procedure is a student t value from a distribution with $\underline{n_1 + n_2 - 2}$ degrees of freedom. The degrees of freedom in the t distribution equal the denominator from the formula for computing Sp.

11-5 Estimating the Difference Between Two Population Means--Paired Samples:

Sometimes, even though we might be confident that the assumptions underlying the two sample procedures from the previous section are met, we will pair observations to (hopefully) control unwanted variation. The paired sample Student t procedure requires <u>dependent</u> samples of the same size. The pairing must be done before the data is collected and reflect some natural or logical grouping. Since the paired sample procedure utilizes small samples, we must be willing to assume that the population of all possible difference measurements is normally distributed.

A typewriter manufacturer has developed a new correcting model which will hopefully help secretaries type faster than they can with a competing model. The following experiment was designed to investigate whether secretaries type faster with the new model.

It is likely that any sample of secretaries chosen to test the new machine will have varied typing speeds. If they do, this unwanted variation (secretaries typing speeds) could make it difficult to determine whether the typing speeds differ for the two machines. Therefore 6 secretaries were randomly selected and checked for typing speed on both models. Half of the secretaries typed first on the new model and then on the competing model and half did the reverse. Now we can focus only on the <u>difference</u> in words per minute typing speeds between models because we have blocked out variation due to the different skill levels of the secretaries. The test data is presented below with summary statistics.

Secretary	New Model	Competing Model	d = Difference
1	81	76	5
2	70	63	7
3	79	74	5
4	75	76	-1
5	80	72	8
6	84	72	12
			36

$$\bar{d} = \frac{36}{6} = 6 \qquad S_d = \sqrt{\frac{\sum_{1}^{6}(d_i - \bar{d})^2}{n-1}} = 4.29$$

If we can assume that the population of difference measurements for all secretaries is normally distributed, then a 95% confidence interval for the true mean difference is

$$\bar{d} \pm (t_{.95,\ 5df}) \frac{S_d}{\sqrt{n}}$$

$$6 \pm (2.571) \frac{(4.29)}{\sqrt{6}}$$

$$6 \pm 4.50$$

or

$$1.50 \text{ ——— } 10.50$$

We estimate that, on the average, secretaries type from 1.5 to 10.5 words per minute faster with the new model.

The key requirement for properly employing the paired-difference procedure is that <u>the pairing must occur by some natural or logical scheme before the data is collected.</u> The pairing is an effort to block out suspected but unwanted variation (e.g., secretaries' typing speeds) so that a possible difference (e.g., typewriters) of primary interest might be more visible.

11-6 <u>Effect of Violating the Assumptions for Small Sample Estimation:</u>

All the procedures presented in the text require random sampling. In addition, for small sample inference we require that the population(s) of interest be normally distributed. If two independent samples are selected, the parent populations must have equal variances (standard deviations).

In general, the small sample procedures presented in Chapter 11 are somewhat insensitive to slight departures from the assumptions of normality and equal variances. That is, if the populations are mound shaped and approximately symmetric (rather than exactly normal), or if compared populations possess variances which differ by less than, say, ten to fifteen percent, these procedures still work very well. Random sampling, however, must <u>always</u> be assured.

```
****************************
*                          *
*         SOLUTIONS        *
*                          *
****************************
```

11-1 In general, the number of degrees of freedom is determined by subtracting from the sample size, the number of statistics that must be calculated to provide an estimate of the variance of the sampling distribution. For example, if our objective is to estimate μ_x, the population mean, we have to compute only S^2, the sample variance to arrive at an estimate for the variance of the sampling distribution. Thus, the appropriate number of degrees of freedom is n - 1.

If our objective is to estimate the difference between two population means, the variance of the sampling distribution is determined by combining S_1^2 and S_2^2. Thus, from the combined sample size ($n_1 + n_2$) we must subtract 2 to get the appropriate degree of freedom.

11-3   (a) The small sample confidence interval formula is

$$\bar{X} \pm t\, S_{\bar{X}}$$

With

$$n = 22 \text{ (degrees of freedom} = n - 1 = 21)$$

for a 95% confidence level $t = 2.080$,

thus the interval is:

$$11.3 \pm 2.08 \frac{2.69}{\sqrt{22}}$$

$$11.3 \pm 1.19$$

$$10.11 \longrightarrow 12.49$$

Thus the investigator is 95% confident that the interval 10.11 hours to 12.49 hours contains the true average daily driving time.

(b) For an 80% confidence level, and degrees of freedom = 21,

$$t = 1.323$$

thus the confidence interval is:

$$11.3 \pm 1.323 \frac{2.69}{\sqrt{22}}$$

$$11.3 \pm .758$$

$$10.54 \text{ hours} \longrightarrow 12.06 \text{ hours}$$

(c) Such factors as whether the sampling procedures used resulted in a representative sample of the population should be considered.

11-5 The small sample confidence interval is:

$$\bar{X} \pm t\, S_{\bar{X}}$$

(a) For $n = 10$ (df = $n - 1 = 9$), and a 98% confidence level,

$$t = 2.821$$

The interval is:

$$5.93 \pm 2.821 \frac{.13}{\sqrt{10}}$$

$$5.93 \pm .12$$

$$5.81 \text{——} 6.05 \text{ oz.}$$

(b) Since the interval of 5.81 oz. to 6.05 oz. contains the value 6 oz., we cannot conclude the advertising is false based upon these data.

(c) If by accuracy we mean that the true average fill is 6 ounces, it is possible that an average of 10 cups could be 5.93 ounces due to the variation in fill that exists. The probability of an average fill of 5.93 ounces or less is found as follows:

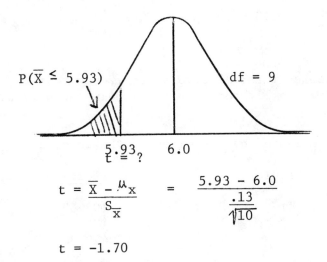

$$t = \frac{\bar{X} - \mu_X}{S_{\bar{X}}} = \frac{5.93 - 6.0}{\frac{.13}{\sqrt{10}}}$$

$$t = -1.70$$

From the t distribution table, the probability in the tail area below $t = -1.70$ is approximately .07. Thus there is about a .07 chance of obtaining a sample mean of 5.93 or less (sample size = 10) from a population with $\mu_X = 6$ and $S_X = .13$.

11-7 This is a common misconception on the part of many individuals who have been exposed only to large sample statistical procedures. Providing the population distributions are approximately normal with equal variances, statistical inferences can be made based upon small samples providing the t distribution is employed.

11-9 We are estimating the difference between two population means. The confidence interval formula is:

$$(\bar{X}_N - \bar{X}_C) \pm t\, S_p \sqrt{\frac{1}{n_N} + \frac{1}{n_C}}$$

where

$$Sp = \sqrt{\frac{S_N^2(n_N - 1) + S_C^2(n_C - 1)}{n_N + n_C - 2}}$$

$$Sp = \sqrt{\frac{.29^2(11 - 1) + .32^2(11-1)}{11 + 11 - 2}}$$

$$Sp = .31$$

We have selected a confidence level of 98%. For a 98% confidence level and $n_N + n_C - 2 = 20$ degrees of freedom, t = 2.528. The desired confidence interval is:

$$(2.14 - 2.27) \pm 2.528(.31)\sqrt{\frac{1}{11} + \frac{1}{11}}$$

$$-.13 \pm .33$$

$$-.46 \longrightarrow .20$$

Since the 98% confidence interval of -0.46 to .20 contains 0, we cannot conclude the new cake mix is lighter than the leading competitor's.

11-11 Precision is measured by the width of the interval.

Sample	Precision
1	193.62
2	173.57
3	429.30

Because sample 2 produced the narrowest interval, it is the most precise estimate of the mean retirement income.

11-13 The confidence interval equation is:

$$\overline{X} \pm t\frac{S_x}{\sqrt{n}}$$

The total width is 173.57. Therefore the tolerable error is 173.57/2 = 86.79. Thus:

$$\overline{X} \pm 86.79$$

and

$$e = t\frac{S_x}{\sqrt{n}}$$

thus

$$86.79 = t\frac{S_x}{\sqrt{n}}$$

Now

For df = n - 1 = 9 and a 95% confidence level t = 2.262. Since Sx = 210 we have:

$$86.79 = 2.262\frac{210}{\sqrt{n}}$$

Solving for n, we get:
$$n = \frac{t^2 S_x^2}{e^2}$$

n = 29.95 = 30

The sample size is about 30.

Note, this problem has a "catch 22" in that we assumed the sample size was n = 10 in determining the t value. However, if the sample size is n = 30 as we have computed, then the t value must be determined for n - 1 = 29 degrees of freedom. Other than using the Z value from the standard normal table, there is no direct way of arriving at the sample size without going through this circular process.

11-15 Each interval provides an estimate of the mean retirement income for people over 65 years of age. Taking into account problems 11, 12, 13, and 14, the first interval would not appear to have any advantages over the other two other than it was less expensive to arrive at because it has the smallest sample size. It has 84% confidence (see problem 12) and the second greatest precision.

Sample two has the greatest precision and has a confidence level equal to the interval for sample three. Its only disadvantage would be that it was developed from the largest sample size which would add to the costs involved.

Sample three has the advantage of having a high confidence level, however, it does have low precision relative to the other two.

The reason that the interval from sample three is so wide is due to the very large sample standard deviation relative to the other two samples.

It appears doubtful that we would get such wide variation in Sx values for three random samples from the same normally distributed populations.

11-17 The reason that we have three different interval estimates is that we have selected three different random samples. We expect sampling error. There is no sure way of guaranteeing an estimate is accurate. However, in this case, we would pick interval three because it has essentially the same mean as the other three and is much wider. But in fact, either this interval will include the true mean or it won't.

11-19 (a) There are many potential problems with the way in which the sampling was performed. First there were no apparent controls to assure that one class couldn't, by chance, have all the best students and would therefore perform better regardless of the teaching method. Other potential problems involve the ability to attach any difference in performance to the two teaching methods.

(b) To test this statistically we must determine an interval estimate of the difference. The interval formula is:

$$(\bar{X}_o - \bar{X}_n) \pm t \, Sp \sqrt{\frac{1}{n_o} + \frac{1}{n_N}}$$

where

$$Sp = \sqrt{\frac{S_o^2 (n_o - 1) + S_N^2 (n_N - 1)}{N_o + n_N - 2}}$$

$$Sp = \sqrt{\frac{7.2^2 (37 - 1) + 6.9^2 (33 - 1)}{37 + 33 - 2}}$$

$$Sp = 7.06$$

with 95% confidence and 68 degrees of freedom

$$t = Z = 1.96$$

The confidence interval is:

$$(78.9 - 81.3) \pm 1.96 \, (7.06) \sqrt{\frac{1}{37} + \frac{1}{33}}$$

$$-2.4 \pm 3.31$$

$$-5.71 \text{ ———— } .91$$

Since the 95% interval of -5.71 to .91 contains 0, Kent cannot conclude there is a difference between the two teaching methods.

11-21 We assume that 10 of each type of platforms are tested.

Our approach will be to develop a confidence interval estimate for the difference between two means and base our conclusion on the results of this estimate.

The confidence interval formula is:

$$(\bar{X}_A - \bar{X}_B) \pm t\, S_p \sqrt{\frac{1}{n_A} + \frac{1}{n_B}}$$

where

$$S_p = \sqrt{\frac{S_A^2(n_A - 1) + S_B^2(n_B - 1)}{n_A + n_B - 2}}$$

$$S_p = \sqrt{\frac{.6^2(10 - 1) + .48^2(10 - 1)}{10 + 10 - 2}}$$

$$S_p = .543$$

Choosing a 90% confidence level, $t = 1.734$

The interval is:

$$(5.2 - 4.9) \pm 1.734\,(.543)\sqrt{\frac{1}{10} + \frac{1}{10}}$$

$$.3 \pm .42$$

$$-.12 \text{ to } .72$$

Since the 90% confidence interval of $-.12$ to $.72$ contains 0, these data give no reason to prefer either manufacturer based on average assembly time.

11-23 The confidence interval estimate for the average contribution takes the following form:

$$\bar{X} \pm t\, S_{\bar{X}}$$

where

$$\bar{X} = 88.00$$

$$S_{\bar{X}} = \frac{S_x}{\sqrt{n}}$$

$$= \frac{25.00}{520}$$

$$= 5.59$$

For a 95% confidence interval and 19 degrees of freedom

$$t = 2.093$$

then the confidence interval is:

$$88.00 \pm 2.093 \, (5.59)$$

$$88.00 \pm 11.70$$

$$76.30 \text{ ------ } 99.70$$

To meet the goal, the individual contributions will have to average $100 each. Since the interval we found does not contain $100, at the 95% level it appears they will not raise $1,500,000. The treasurer's report, based upon these data, should not be optimistic.

11-25 The manager overseeing the study has required that programming time measurements be collected from the same programmer under each environment, batch and on-line. Thus he is attempting to control for the possible differences in ability between the worker to avoid the possible random assignment of all fast workers to one environment and slower workers to the other, thereby disturbing the results.

Consequently, a paired sample confidence interval is desirable:

$$\bar{d} \pm t \, \frac{S_d}{\sqrt{n}}$$

where:

$\bar{d}$ = average paired difference

$$\bar{d} = \frac{\sum d}{n}$$

S_d = standard deviation of the paired differences

$$S_d = \sqrt{\frac{\sum (d - \bar{d})^2}{n-1}}$$

The differences are:

Programmer	Batch	On-line	d	$(d-\bar{d})^2$
1	9.8 hrs.	8.5 hrs.	1.3 hrs.	.36
2	11.7	10.0	1.7	1.0
3	5.6	5.8	-.2	.81
4	10.5	8.6	1.9	1.44
5	14.3	13.0	1.3	.36
6	9.2	11.0	-1.8	6.25
			$\bar{d} = 4.2/6$	10.22
			$\bar{d} = .7$	

Then:

$$\bar{d} = .7$$

$$Sd = \sqrt{\frac{10.22}{5}} = 1.43$$

For a 90% confidence interval and 5 degrees of freedom from the t-distribution

$$t = 2.015$$

The desired confidence interval is

$$\bar{d} \pm t \frac{Sd}{\sqrt{n}}$$

$$.7 \pm 2.015 \frac{1.43}{\sqrt{6}}$$

$$.7 \pm 1.176$$

$$-.476 \text{ ——— } 1.876$$

This confidence interval includes a difference of 0.0. Thus, based upon these sample data, we conclude that there possibly is no difference between the average programming time between the batch and on-line environments.

11-27 The 95% confidence level interval is wider (less precision) than the 90% interval. However, both intervals are centered at the same point, $\bar{d}$.

CHAPTER 12

INTRODUCTION TO HYPOTHESIS TESTING

The statistical estimation techniques presented in Chapters 10 and 11 are most appropriate if we have little or no idea of the numerical value of the parameter of interest. Interval estimation helps us develop a range of possible values for the parameter.

Hypothesis testing is most appropriate if we have a suggested value for the parameter. Hypothesis testing provides a method for judging sample information to decide whether it supports or refutes the suggested value for the parameter.

12.1   Reasons for Testing Hypotheses

Statistical hypothesis testing is an analytical decision making aid which allows the user to control the probability of committing an error. For most hypothesis testing applications we can specify, in advance, the maximum tolerable chances of committing errors, and thus insure that our decision making procedure leads us to logical results.

12.2   The Hypothesis Testing Process

All hypothesis tests have very much the same format. We must first formulate two conflicting descriptions of the world. These two positions are called the null hypothesis and the alternate hypothesis. Between them they include every possible state of the world with respect to the parameter of interest. Therefore, either the null hypothesis is true or the alternate hypothesis is true.

Second, in anticipation of collecting sample information, we must decide what type of sample values lend support to the null hypothesis and what type of sample values refute the null hypothesis and support the alternate hypothesis. That is, we must establish a decision rule which specifies whether, upon viewing the sample information, we will accept the null hypothesis or reject it and accept the alternate hypothesis.

Third, we must search diligently for sample information using a specified sampling design. The sampling design must insure that we obtain a random sample from the population of interest.

Fourth, we must compare the sample information to the decision rule and reach a conclusion about the parameter of interest. This conclusion should be stated in the specific circumstances of the problem and ought to lead logically to one and only one action by the decision maker.

Some guidelines for establishing hypotheses are:

1. The null hypothesis must contain some form of an _equality_.

2. Hypotheses should be stated in terms of parameters whenever possible. Hypotheses <u>never</u> include statements about statistics.

3. Often the situation we are most interested in detecting, if it occurs, is placed in the alternate hypothesis.

4. The null and alternate hypotheses must be mutually exclusive and, collectively, they must include every possible value for the parameter of interest.

We can represent every possible action and state combination for an hypothesis test in the decision matrix below.

		State of the World	
		H_o True	H_o False
Action by Decision Maker	Accept H_o	Correct Decision	Type II Error
	Reject H_o	Type I Error	Correct Decision

The decision rule is formulated to directly control the probability of committing a Type I Error. The probability of committing a Type I Error is called alpha and is represented as α. A well designed hypothesis test also provides for control of the probability of committing a Type II Error. The probability of committing a Type II Error is called beta and is represented by β. For most hypothesis tests beta is more difficult than alpha to control at a specified level. However, in any decision situation we should recognize that errors are possible and strive to keep the chances of committing errors as low as possible.

Hypothesis testing works on the following rationale. Sample information which, probabilistically, is likely to be observed if the null hypothesis is true, supports the null hypothesis. Therefore, if the sample values are close to the hypothesized value of the parameter, we conclude that the null hypothesis could be true. Sample information which, probabilistically, is unlikely if the null hypothesis is true refutes the null hypothesis. Therefore, if the sample values differ considerably from the hypothesized value of the parameter, we conclude that the null hypothesis must be false.

12.3 One-Tailed Hypothesis Tests

A one-tailed hypothesis test is appropriate if we are most concerned about possible deviations from the hypothesized parameter value in only one direction. The null and alternate hypothesis take one of the following forms.

1: H_o: parameter $\leq$ specified value

H_a: parameter $>$ specified value

OR

2: H_o: parameter $\geq$ specified value

H_a: parameter $<$ specified value

Hypothesis set 1 leads to a one-sided decision rule in the right hand tail of the critical distribution. That is, we only reject H_o in favor of H_a if we observe sample values much larger than the hypothesized value of the parameter.

Hypothesis set 2 leads to a one-sided decision rule in the left hand tail of the critical distribution. That is, we only reject H_o in favor of H_a if we observe sample values much smaller than the hypothesized value of the parameter.

For one-tailed hypothesis tests all of the probability of a Type I Error (alpha) is placed in one tail of the critical distribution.

12.4 Two-Tailed Hypothesis Tests

In a two-tailed hypothesis test we are equally concerned about deviations from the hypothesized parameter value in either direction. The null and alternate hypothesis take the form:

1: H_o: parameter $=$ specified value

H_a: parameter $\neq$ specified value

This hypothesis set leads to a decision rule which has half of the probability of a Type I Error in each tail of the critical distribution. Thus we would reject H_o in favor of H_a if we observe sample values either much smaller or much larger than the hypothesized parameter.

12.5 Hypothesis Testing-Population Standard Deviation Unknown, Large Sample

If the population standard deviation is unknown, as is usually the case, the critical value A must be established using the sample standard deviation. For large sample ($n > 30$) tests we find A by solving:

$$z = \frac{A - \mu_x}{s_x/\sqrt{n}}$$

Yielding:

$$A = \mu_x + z\,(S_{x/\sqrt{n}}).$$

Remember that z takes on a positive sign for A values in the right hand tail and a negative sign for A values in the left hand tail. Therefore, in two-tailed tests:

$$A_L = \mu_x - z\,(S_{x/\sqrt{n}})$$

and

$$A_H = \mu_x + z\,(S_{x/\sqrt{n}}).$$

12.6 Type II Errors and Power of a Hypothesis Test

Notice that all of the decision rules for hypothesis tests directly control alpha, the maximum tolerable probability of a Type I Error. The tolerable probability of committing a Type I Error should be selected only after considering the relative costs of committing either a Type I Error or a Type II Error. If a Type I Error would be expensive relative to a Type II Error we should select alpha small. A small alpha lessens the chance of rejecting a true null hypothesis. If a Type II Error would be relatively more expensive, we should select alpha larger. Selecting alpha larger creates a smaller acceptance region in the decision rule and thus lessens the chance of accepting a false null hypothesis (Type II Error).

Therefore, in general, the chances of committing Type I and Type II Errors are inversely related. We can select alpha smaller to protect against a Type I Error, but we must recognize that beta will increase if we do. Conversely, if we select a larger alpha then beta will be relatively smaller.

A well designed hypothesis test allows the decision maker to control both alpha and beta at tolerable levels. Usually this is accomplished by selecting a large sample. Logically, as we get more information (increase the sample size) we should be less likely to commit an error of either kind. Recall, however, that sampling can be expensive and increasing the sample size may not be economically feasible.

SOLUTIONS

12.1 $H_o: \mu \leq 20$ minutes

$H_A: \mu > 20$ minutes

The hypotheses have been stated in this manner to be in agreement with the goal of trimming a tree in no more than 20 minutes. If the null hypothesis is accepted, the Balock Company will conclude that they are within their goal on the average. If they reject the null hypothesis, they must conclude that they are exceeding their goal.

12.3 A Type II error occurs when a <u>false</u> null hypothesis is <u>accepted</u>. The chances of making a Type II error can be reduced by <u>increasing</u> alpha or <u>increasing</u> the sample size or both.

12.5 I might hypothesize that the coin is fair. Given this hypothesis, the chances of 20 consecutive tails is:

$$(1/2)^{20} = .00000095$$

The logic of hypothesis testing says to reject the null hypothesis if the probability of the observed event (20 tails) is <u>too small</u> given the null hypothesis of a fair coin. The question is: Is .00000095 <u>too small</u>? If the probability of the observed result is less than the specified alpha, the answer is yes.

12.7 For both hypotheses, a socially responsible company would set β smaller than α. The extent of the difference between the two depends upon how large a sample size can be taken. The larger the sample the more possible it is to control α and β as desired.

12.9 Under conditions of desired expansion, the company would no doubt wish to increase its customer acceptance rate. Thus, they would be more inclined to accept customers even if this means accepting poor risks.

The company would no doubt make the acceptance policies more liberal and thus cut down on the Type I error probability. In doing so they could run a greater risk of committing a Type II error.

12.11 (a) H_o: $\mu_x \geq 3.4$ pages

H_A: $\mu_x < 3.4$ pages

(b) Hypothesized Distribution

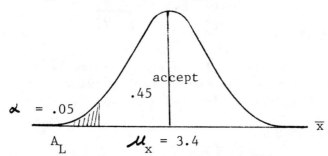

From normal table

$z = -1.645$ which means A_L is 1.645 standard deviations from μ_x.

Then solving for A_L:

$$z = \frac{A_L - \mu_x}{\sigma_{\bar{x}}} = \frac{A_L - \mu_x}{\sigma_x/\sqrt{n}}$$

$$A_L = \mu_x + z\, \sigma_{\bar{x}}$$

$$A_L = 3.4 + (-1.645)\frac{(.5)}{\sqrt{400}}$$

$$A_L = 3.359$$

Since $\bar{x} = 3.2 < 3.349$ pages per hour, the null hypothesis should be <u>rejected</u>. The sample results fall in the rejection region.

The Write-Type Company, based on this hypothesis test, should elect not to make the claim that their typists average at least 3.4 pages per hour. Either a lower claim is needed or faster typists need to be hired. However, a Type I error may have been committed.

12.13 H_o: $\mu_x \geq 3.4$
H_A: $\mu_x < 3.4$
$n = 300$
$\bar{x} = 3.34$

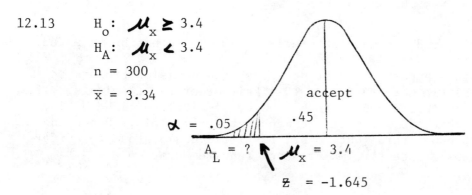

Solving for A_L:
$$z = \frac{A_L - \mu_x}{\sigma_{\bar{x}}}$$

$$A_L = \mu_x + z\,\sigma_{\bar{x}}$$

$$A_L = 3.4 + (-1.645)\frac{.5}{\sqrt{300}}$$

$$A_L = 3.35$$

Since $\bar{x} = 3.34 < 3.35$ the decision should be to reject the null hypothesis. Again this means that the secretarial service is not justified in making the claim that they type at least 3.4 pages per hour.

12.15 (a) $H_o: \mu_x \geq 50$ hours

$H_A: \mu_x < 50$ hours

Note, we have set this up as a one tailed hypothesis test.

(b) $\sigma_x = 25$

$n = 100$

$$\sigma_{\bar{x}} = \frac{\sigma_x}{\sqrt{n}} = \frac{25}{\sqrt{100}} = 2.5$$

$\alpha = .10 \quad .40 \quad$ accept

$A_L = ? \quad 50$

$z_{.40} = -1.28$

Then solving for A_L:

$$z = \frac{A_L - \mu_x}{\sigma_{\bar{x}}}$$

$$A_L = \mu_x + z\,\sigma_{\bar{x}}$$

$$A_L = 50 + (-1.28)(2.5)$$

$$A_L = 46.8$$

Since $48 > 46.8$ the manufacturer's claim should <u>not</u> be rejected.

(c) If this hypothesis test dominates the decision process in this case, the Bo-Little Corporation would likely order the terminals. If, however, the null hypothesis proves to be false, Bo-Little will be buying computer terminals which do not meet the manufacturer's claim. Depending upon how poor the product turns out to be, Bo-Little could stand to suffer financial losses due to excessive maintenance requirements.

12.17 $H_O: \mu_x \geq 84$

$H_A: \mu_x < 84$

$\alpha = .05$

$n = 100$

$\bar{x} = 82$

$S_x = 10$

$z_{.45} = -1.645$ from normal distribution table

Solve for A_L:

$$z = \frac{A_L - \mu_x}{\sigma_{\bar{x}}}$$

$$A_L = \mu_x + z\sigma_{\bar{x}}$$

$$A_L = 84 + (-1.645)\frac{10}{\sqrt{100}}$$

$$A_L = 82.355$$

Since $\bar{x} = 82 < 82.355$ <u>reject</u> H_O and conclude that production is operating at less than the standard.

The production manager might want to do a follow-up study to determine why the output level has fallen.

12.19   The objective is to determine the probability of accepting a false hypothesis. This value we are trying to find is called Beta.

We solve for Beta as follows:

$H_O: \mu_x \geq 84$

$H_A: \mu_x < 84$

$S_x = 10$

$n = 100$

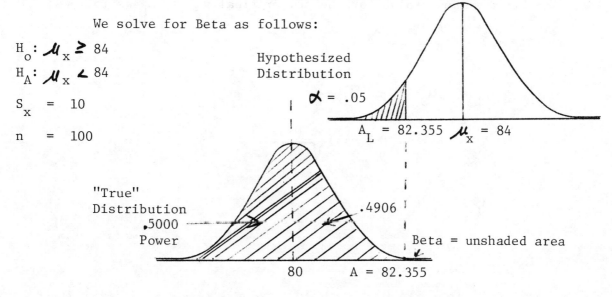

To find Beta:

$$z = \frac{A - \mu_x}{\sigma_{\bar{x}}}$$

$$z = \frac{82.355 - 80.0}{\frac{10}{\sqrt{100}}} = 2.35$$

Area between $z = 2.35$ and μ_x is .4906. Thus, Beta is equal to .5000 − .4906 = .0094.

Thus, the probability of accepting that the average production is at least 84 tables when in fact the true average is only 80 is .0094. The chance is quite small of that happening.

12.21   An increase in the sample size from 100 to 144 will reduce $\sigma_{\bar{x}}$. This will result in a smaller value of Beta. For example:

If the "true" mean is 83 we find Beta as follows:

$H_o: \mu_x \geq 84$

$H_A: \mu_x < 84$

$\alpha = .05$

$S_x = 10$

$\iota = 144$

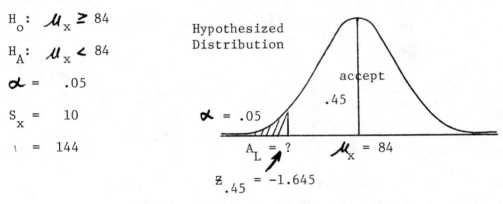

$z_{.45} = -1.645$

First we must solve for the new critical value, A_L as follows:

$$z = \frac{A_L - \mu_x}{\sigma_{\bar{x}}}$$

$$A_L = \mu_x + z\, \sigma_{\bar{x}}$$

$$A_L = 84 + (-1.645)\frac{10}{\sqrt{144}}$$

$$A_L = 82.63$$

Now:

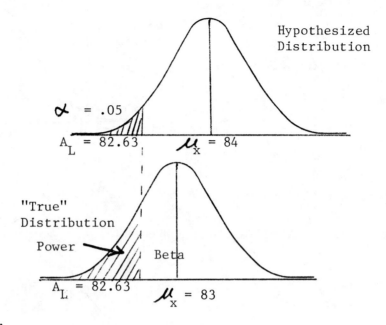

To find Beta:

$$A = \frac{A - \mu_x}{\sigma_{\bar{x}}}$$

$$Z = \frac{82.63 - 83.0}{\frac{10}{\sqrt{144}}}$$

$$Z = -.44$$

Area between $Z = -.44$ and μ_x is .1700.

Beta = .5000 + .1700 = .6700

We see that the increase in sample size from 100 to 144 has resulted in a decrease in Beta from .7389 to .6700. Obviously a decision maker would have to weigh the cost of the increased sample size against the benefit of reducing the Type II error probability.

12.23 We are asked to calculate Beta for the same situation with the exception that the sample size has been increased from 49 to 64.

We can do this as follows:

H_o: $\mu_x \geq 0.0$

H_A: $\mu_x < 0.0$

n = 64

σ_x = \$4

A_L = -\$2.0

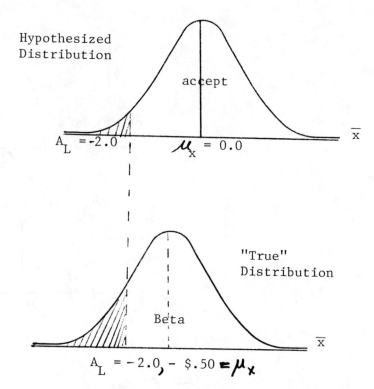

To find Beta:

$$Z = \frac{A_L - \mu_x}{\sigma_{\bar{x}}}$$

$$Z = \frac{-\$2.0 - (-\$1.50)}{\frac{4}{\sqrt{64}}} = -1.0$$

Area between $Z = -1.0$ and μ_x is equal to .3413.

Thus Beta = .3413 + .5000 = .8413.

12.25 The appropriate null and alternative hypotheses are:

H_o: $\mu_x \leq 100.00$

H_A: $\mu_x > 100$

Where μ_x = average tax owed by individual taxpayer. The Decision Rules for α = .10 is:

Take a sample of n = 40 and determine $\bar{x}$:

if $\bar{x} \leq A$, accept H_o
if $\bar{x} > A$, reject H_o

The critical value, A, is determined as follows:

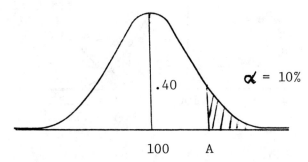

$Z = 1.28$ From the normal table

We are given $\sigma_x^2 = 2400$
$\sigma_x = 48.99$

Then:

Using $Z = \dfrac{A - \mu_x}{\sigma_{\bar{x}}}$

$1.28 = \dfrac{A - 100}{\dfrac{48.99}{\sqrt{40}}}$

$A = 100 + 1.28 \dfrac{(48.99)}{\sqrt{40}}$

$= 109.91$

Then:

If $\bar{x} > 109.91$, reject H_o, otherwise do not reject.

Since $\bar{x} = 114.00$, which is greater than A, the decision rule leads us to reject H_o. The sample data refutes the director's claim.

12.27 We must determine the value of Beta if the true value of μ_x is $115. We start with:

$H_o: \mu_x \leq 100.00$

$H_A: \mu_x > 100.00$

$\alpha = .10$

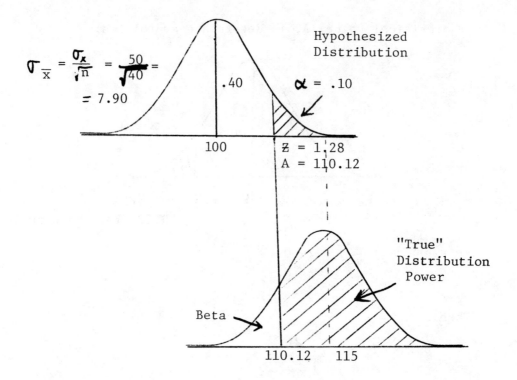

First solve for A:

$$z = \frac{A - \mu_x}{\frac{\sigma}{\sqrt{n}}}$$

Then:

$$A = z\frac{\sigma}{\sqrt{n}} + \mu_x$$

$$A = (1.28)\,7.90 + 100$$

$$A = 110.12$$

To find Beta:

$$z = \frac{110.12 - 115}{7.90} = -.62$$

From normal table, area corresponding to $z = -.62$ is .2324. Therefore, to find Beta we subtract .2324 from .5000.

$$\text{Beta} = .5000 - .2324$$

$$= .2676$$

12.29   We need to determine the value of Beta when the true value of μ_x = 13100. We start with:

We will assume the population standard deviation is 6000.

H_o: $\mu_x \leq \$13,000$

H_A: $\mu_x > \$13,000$

$\alpha = .05$

$S_{\bar{x}} = \dfrac{S_x}{\sqrt{n}} = \dfrac{6000}{\sqrt{36}} = 1000$

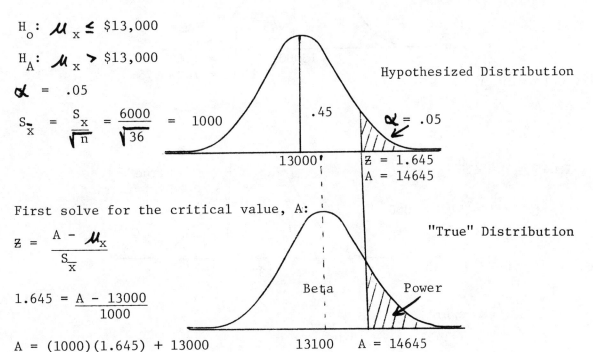

First solve for the critical value, A:

$Z = \dfrac{A - \mu_x}{S_{\bar{x}}}$

$1.645 = \dfrac{A - 13000}{1000}$

$A = (1000)(1.645) + 13000$

$A = \$14,045$

To find Beta:

$Z = \dfrac{14645 - 13100}{1000} = \dfrac{1545}{1000} = 1.55$

Area from normal table corresponding to $Z = 1.55$ is .4394

Thus Beta = .5000 + .4394 = .9394

12.31   Our objective is to determine the power of the test when μ_x = 13200 and sample size is A = 64. We start with:

H_o: $\mu_x \leq 13000$

H_A $\mu_x > 13000$

$\alpha = .05$

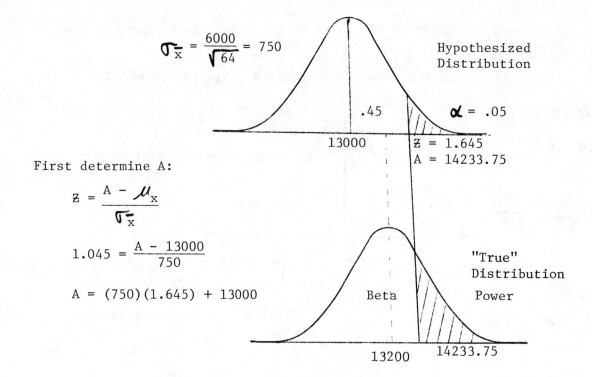

First determine A:

$$z = \frac{A - \mu_x}{\sigma_{\bar{x}}}$$

$$1.045 = \frac{A - 13000}{750}$$

$$A = (750)(1.645) + 13000$$

To find Power:

$$z = \frac{14233.75 - 13200}{750}$$

$$z = 1.38$$

From the normal table, area corresponding to $z = 1.38$ is .4162. Thus, Power = .5 − .4162 = .0838. Thus in contrast to problem 30 which had sample size = 36, a sample size of n = 64 improved power slightly from .0735 to .0838.

CHAPTER 13

ADDITIONAL TOPICS IN HYPOTHESIS TESTING

In Chapter 12 we introduced hypothesis testing concepts using a single population mean as a focal point. In this chapter we consider tests about a single population proportion, tests comparing two population means, tests comparing two population proportions, and tests concerning both one population variance and two population variances.

The size of the sample(s) is also an important issue in hypothesis testing. In this chapter we consider the problem of choosing the sample size necessary to control the probability of both Type I Errors and Type II Errors.

We also present procedures which are appropriate if the sample size is small ($n \leq 30$) and tests based on the standard normal cannot be used.

13.1 Controlling Type I and Type II Errors

The probability of committing Type I and Type II Errors can be controlled simultaneously by solving for an adequate sample size. To control the probability of a Type I Error at level alpha if the true mean is μ_x and the probability of a Type II Error at level beta if the true mean is μ_a, we need the sample size determined by

$$n = \left[\frac{(z_\alpha - z_\beta) S_x}{\mu_x - \mu_a} \right]^2$$

Where:

z_α = the standard normal score needed to control alpha

z_β = the standard normal score needed to control beta if the true mean is μ_a

μ_x = the value of the mean specified in the null hypothesis

μ_a = the alternate specified value for the mean

S_x = the best available estimate of the population standard deviation.

For the example in the text:

z_α = -2.07

z_β = 1.65

$$\mu_x = 1200$$
$$\mu_a = 1180$$
$$S_x = 40$$

So:
$$n = \left[\frac{(-2.07 - 1.65)\,40}{1200 - 1180}\right]^2$$

$$n = 55.35 \text{ or } 56.$$

This formula for sample size can be verified by solving the simultaneous equations represented by formulas 13.1 and 13.2 in the text.

13.2 Hypothesis Tests About A Population Proportion

The general format for hypothesis tests concerning a population proportion must conform to that presented for a single mean in Chapter 12. First, establish the null and alternate hypothesis. Then determine a decision rule to control the probability of a Type I Error at level alpha. Use a critical value

$$A = p + Z\sqrt{\frac{p(1-p)}{n}}$$

Where: p = hypothesized proportion
n = sample size
Z = standard normal score

Compare the observed sample proportion $\hat{p}$ to A and formulate a conclusion by applying the decision rule. Remember that Z is positive for values of A in the right hand tail and negative for values of A in the left hand tail.

13.3 Hypothesis Testing About the Difference Between Two Population Means:

Hypotheses concerning the difference between two means have the following general form

$$H_o: \mu_{x_1} - \mu_{x_2} = \text{specified value}$$

against

$$H_a: \mu_{x_1} - \mu_{x_2} \neq \text{specified value}$$

The specified value (hypothesized difference) is often zero (no difference in means) and one-sided tests in either direction are also possible.

The critical value, A, is determined using the general formula of

$$A = \text{hypothesized difference} + Z \cdot (\text{standard error})$$

The standard error is $\sqrt{\dfrac{\sigma_{x_1}^2}{n_1} + \dfrac{\sigma_{x_2}^2}{n_2}}$

if the population variances are known. If the population variances are not known we substitute the sample variances in the standard error. The standard normal score, Z, is positive for values of A in the right hand tail and negative for values of A in the left hand tail.

13.4   Hypothesis Testing About the Difference Between Two Population Proportions:

Hypotheses concerning the difference between two proportions have the following general form

$$H_o: \quad p_1 - p_2 = \text{specified value}$$

against

$$H_a: \quad p_1 - p_2 \neq \text{specified value}.$$

The specified value (hypothesized difference) is often zero (no difference in proportions) and one-sided tests in either direction are also possible.

The critical value, A, is determined using the general formula of $A = \text{hypothesized difference} + Z \cdot (\text{standard error})$. As long as we hypothesize a difference of zero (specified value = 0) the standard error should be computed using

$$S_{p_1 - p_2} = \sqrt{\bar{p}(1-\bar{p})\left(\frac{1}{n_1} + \frac{1}{n_2}\right)}$$

where $\bar{p}$ is a weighted combination of the sample proportions defined by

$$\bar{p} = \frac{n_1 \hat{p}_1 + n_2 \hat{p}_2}{n_1 + n_2}$$

The standard normal score, Z, is positive for critical values in the right hand tail and negative for critical values in the left hand tail.

13.5 Hypothesis Testing With Small Samples and σ_x Unknown:

If we are restricted to small sample sizes (rule of thumb $n \leq 30$) and the population standard deviation is unknown the standard normal distribution cannot be used as the foundation for hypothesis tests. However, if we can assume that the sampled population(s) is (are) normal then the Students t distribution may be used.

Single Population Mean

The format of this test procedure is identical to that presented in Chapter 12. The only operational difference occurs in the calculation of the critical value A. Now

$$A = \text{hypothesized parameter} + t \cdot (\text{standard error})$$

where t is a students t value with n-1 degrees of freedom chosen to control the probability of a Type I Error at level alpha. The standard error must be computed from sample data and equals

$$S_x/\sqrt{n}.$$

Difference Between Two Means

The requirements to employ this test procedure are the same as for the small sample, difference between means confidence interval in Chapter 11. Again we should use a pooled estimate of the (assumed) common population variance and therefore

$$A = \text{hypothesized difference} + t \cdot S_{pooled} \sqrt{\frac{1}{n_1} + \frac{1}{n_2}}$$

where t has $(n_1 + n_2 - 2)$ degrees of freedom. Remember that t is negative for A values in the left hand tail and positive for A values in the right hand tail.

13.6 An Alternative Way to Test Hypotheses:

In all of the hypothesis tests thus far we have computed critical values of A for use in the decision rule. Then we compared the observed sample information to A and expressed the appropriate conclusion.

In every one of the previous hypothesis tests we could have expressed the decision rule in terms of standard normal scores or students t scores, if appropriate. Then we transform the observed sample value to the standard normal or students t distribution to apply the decision rule.

Decision rules expressed in terms of the sampling distribution for the statistic (A values) and decision rules formulated using the standard normal or Students t (when appropriate) will always lead to consistent conclusions. This is true because the values of A and the critical Z or t values must be established to isolate the same area in the tail(s) of their respective distributions. Thus Type I Error probabilities must be equal.

The calculated test statistics for decision rules formulated using standard normal scores or Students t scores have the same general form. The general form is

$$\text{test statistic} = \frac{\text{observed sample value} - \text{hypothesized value}}{\text{standard error}}$$

13.7 Some Other Hypothesis Tests

The tests introduced thus far have all focused on the location parameter. We may also wish to make inferences about the dispersion parameter for a population.

Single Population Variance

Hypothesis tests about the value of a population variance typically have the following form:

$$H_o: \sigma_x^2 \leq \text{specified value}$$

$$H_a: \sigma_x^2 > \text{specified value}$$

A test statistic is computed using:

$$x^2 = \frac{(n-1) s_x^2}{\sigma_x^2}$$

where

$$x^2 = \text{test statistic}$$
$$n = \text{sample size}$$
$$s_x^2 = \text{sample variance}$$
$$\sigma_x^2 = \text{hypothesized variance (specified value)}$$

The calculated test statistic is compared to a critical chi-square value which isolates alpha in the right hand tail of a chi-square distribution with (n-1) degrees of freedom. We should reject the hypothesis if the calculated test statistic exceeds the critical value from the table.

Two Population Variances

A typical set of hypotheses about two population variances is

$$H_o: \sigma_1^2 = \sigma_2^2$$

$$H_a: \sigma_1^2 \neq \sigma_2^2$$

We can examine this set of hypotheses by calculating a test statistic defined as

$$F_{cal} = \frac{S_1^2}{S_2^2}$$

where S_1^2 is the larger sample variance and S_2^2 is the smaller sample variance. We then compare this calculated F to a critical F-distribution with $D_1 = n_1 - 1$ degrees of freedom in the numerator and $D_2 = n_2 - 1$ degrees of freedom in the denominator.

Since the alternate hypothesis contains a "not equal to" sign we should logically construct a two tailed decision rule. However, since we structured our test statistic with the larger sample variance in the numerator we should only reject H_o if F_{cal} exceeds the critical value from the table.

Whenever we have a two-sided alternate hypothesis we want to select a tabled value which isolates one-half of alpha in the right hand tail. Then, because of the structured test statistic, we will have the true probability of a Type I Error equal to alpha.

SOLUTIONS

13.1 A Type I error results when the sample statistic leads to rejection of a _true_ null hypothesis.

An example of a serious or costly Type I error can be observed by considering the production design engineer at a major automobile manufacturing company who hypothesizes that a certain assembly line configuration will produce an average of 500 cars a day or more. If he rejects this hypothesis based upon a sample of days' output and is wrong, he may well end up changing the assembly line system needlessly. This would be very costly.

13.3 Your response, while your own, should show an understanding of the difference between the two types of errors and that the errors result because of sampling error.

13.5 The factors to take into account when considering how large alpha and beta should be involve the costs of making the errors. These costs can be either dollar costs or social costs or both. The greater the costs of committing these errors, the smaller should be the probabilities of committing them. Of course, to make both alpha and beta small can require a "large" sample size which also costs money and time to acquire. The decision maker will have to balance these costs.

13.7 This problem introduces a tighter constraint on the Type II error chances.

The first equation for the critical value, A, remains unchanged.

$$A = 8100 + (-1.45)\frac{1900}{\sqrt{n}}$$

Now, however, dealing with the new constraint, we must solve for A with a new equation.

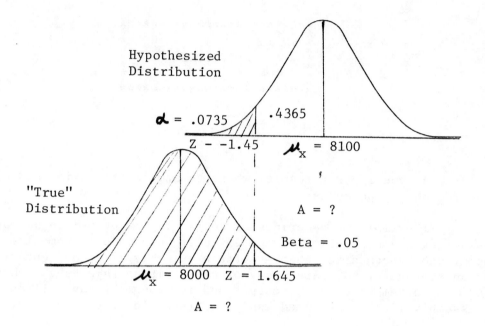

Then solving for A:

$$Z = \frac{A - \mu_x}{\sigma_{\bar{x}}}$$

$$1.645 = \frac{A - 8000}{\frac{1900}{\sqrt{n}}}$$

$$A = 8000 + 1.645 \frac{1900}{\sqrt{n}}$$

We have two equations for A both containing the unknown sample size, n. Now, to find the optimal sample size, we set the equations for A equal to each other and solve for n:

$$A = 8100 + (-1.45) \frac{1900}{\sqrt{n}}$$

$$A = 8000 + 1.645 \frac{1900}{\sqrt{n}}$$

Then:

$$8100 + (-1.45) \frac{1900}{\sqrt{n}} = 8000 + 1.645 \frac{1900}{\sqrt{n}}$$

144

$$100 + (-1.45)\frac{1900}{\sqrt{n}} = 1.645\frac{1900}{\sqrt{n}}$$

Multiply both sides by $\sqrt{n}$:

$$100\sqrt{n} + (-1.45)(1900) = 1.645(1900)$$

$$100\sqrt{n} = 5880.5$$

$$n = 3459$$

Optimal sample size has been increased from 2690 to 3459.

The new critical value is found:

$$A = 8000 + 1.645\frac{1900}{\sqrt{3459}}$$

$$A = 8053.14$$

Thus, optimal Decision Rule is:

Select sample of 3459. If $\bar{x} < \$8053.14$ we should reject the null hypothesis that $\mu_x = \$8100$.

The optimal sample size was increased because the total allowable probability of error was reduced.

13.9 This is a hypothesis test involving proportions.

H_o: $p \geq .90$

H_A: $p < .90$

$\alpha = .10$

$n = 100$

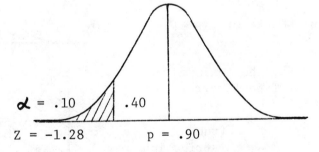

The first step in testing this hypothesis is to determine the critical value, A.

$$Z = \frac{A - p}{\sigma_p}$$

145

To find σ_p we will not use the finite correction since the population of shrubs is large (and unknown) in this problem.

$$\sigma_p = \sqrt{\frac{p(1-p)}{n}}$$

$$\sigma_p = \sqrt{\frac{(.90)(.10)}{100}}$$

$$\sigma_p = .03$$

Now solve for A:

$$-1.28 = \frac{A - .90}{.03}$$

$$A = .90 + (-1.28)(.03)$$

$$A = .8616$$

Decision Rule:

If $\hat{p} < .8616$ reject the null hypothesis.

Since $\hat{p} = \frac{85}{100} = .85 < .8616$

13.11 To hold Beta at half of .7157 or .3578 will require an increase in sample size. We solve for this required sample size as follows:

H_o: $p \geq .90$

H_A: $p < .90$

Constraints:

1. Maximum chance of Type I Error = .10.

2. If p = .88, the maximum chance of Type II error is .3578.

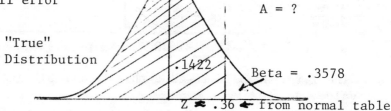

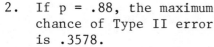

We now set up equations for A as follows:

Hypothesized $\qquad\qquad\qquad\qquad$ "True"

$$Z = \frac{A - p}{\sigma_p} \qquad\qquad\qquad Z = \frac{A - p}{\sigma_p}$$

$$-1.28 = \frac{A - .90}{\sqrt{\frac{(.90)(.10)}{n}}} \qquad\qquad .36 = \frac{A - p}{\sqrt{\frac{(.88)(.12)}{100}}}$$

$$A = .90 + (-1.28)\sqrt{\frac{(.90)(.10)}{n}} \qquad A = .88 + (.36)\sqrt{\frac{(.88)(.12)}{n}}$$

Then set the two equations equal to each other and solve for n.

$$.90 + (-1.28)\sqrt{\frac{(.90)(.10)}{n}} = .88 + (.36)\sqrt{\frac{(.88)(.12)}{n}}$$

$$.02 + (-1.28)\sqrt{\frac{(.90)(.10)}{n}} = (.36)\sqrt{\frac{(.88)(.12)}{n}}$$

Multiply both sides by $\sqrt{n}$:

$$(.02)\sqrt{n} + (-1.28)\sqrt{(.90)(.10)} = (.36)\sqrt{(.88)(.12)}$$

$$.02\sqrt{n} = (.36)\sqrt{(.88)(.12)} + (1.28)\sqrt{(.90)(.10)}$$

$$n = \frac{[(.36)\sqrt{(.88)(.12)} + (1.28)\sqrt{(.9)(.10)}]^2}{(.02)^2}$$

$$n = 627.4$$

$$n = 628 = \text{optimal sample size to meet the constraints.}$$

13.13   A Type II error would take place if the manager accepted that at least 90 percent of the shrubs would live when in fact fewer than 90 percent will live. To make this error would cause a financial hardship on Green Thumb since they would end up replacing a greater number of shrubs than anticipated.

The error (Type I or Type II) which is most serious would depend upon Green Thumb's financial position. If by not selling this shrub with the guarantee means losing sales there will be an opportunity cost. However, if the shrubs are guaranteed and more than expected need to be replaced, the company will incur actual out-of-pocket costs.

13.15

$H_o: \mu_1 - \mu_2 = 0$

$H_A: \mu_1 - \mu_2 \neq 0$

$\alpha = .04$

$\sigma_1^2 = 11.2$

$\sigma_2^2 = 9.6$

$n_1 = 100$

$n_2 = 100$

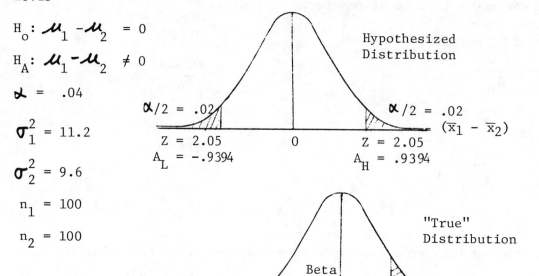

To find Beta:

$$Z = \frac{A_L - (\mu_1 - \mu_2)}{\sqrt{\frac{\sigma_1^2}{n_1} + \frac{\sigma_2^2}{n_2}}}$$

$$A_L = \mu_1 - \mu_2 + Z \sqrt{\frac{\sigma_1^2}{n_1} + \frac{\sigma_2^2}{n_2}}$$

$$A_L = 0 + (-2.05) \sqrt{\frac{11.2}{100} + \frac{9.6}{100}}$$

$$A_L = -.9394$$

Likewise:

$$A_H = .9394$$

To get Beta:

$$Z = \frac{.9394 - .5000}{\sqrt{\frac{11.2}{100} + \frac{9.6}{100}}}$$

$$Z = .96$$

Area between Z = .96 and the assumed true values of $\mu_1 - \mu_2$ is .3315 as found in the normal distribution table:

$$Z = \frac{-.9394 - .5000}{\sqrt{\frac{11.2}{100} + \frac{9.6}{100}}}$$

$$Z = -3.16$$

Area between Z = -3.16 and assumed true value of $\mu_1 - \mu_2$ is .4992.

Beta is:

$$.3315 + .4992 = .8307$$

There is slightly more than an 83 percent chance of committing a Type II error if the true difference between population means is actually .5000 tons/acre.

13.17   Now we wish to keep Beta at .4153 but let alpha grow to as high as .10. The required sample size is found as follows:

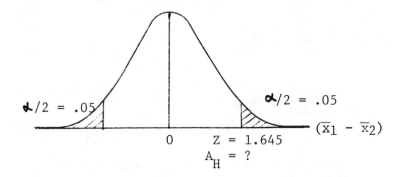

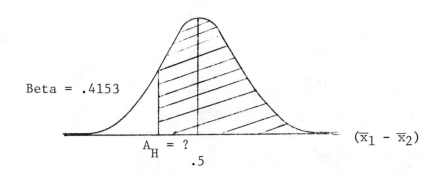

From normal table, Z = -.21

To solve for the common sample size, n:

$$Z = \frac{A_H - 0}{\sqrt{\frac{\sigma_1^2}{n} + \frac{\sigma_2^2}{n}}}$$

$$A_H = 0 + Z\sqrt{\frac{\sigma_1^2}{n} + \frac{\sigma_2^2}{n}}$$

$$A_H = 0 + 1.645\sqrt{\frac{20.8}{n}}$$

And:

$$-Z = \frac{A_H - .5}{\sqrt{\frac{\sigma_1^2}{n} + \frac{\sigma_2^2}{n}}}$$

$$A_H = .5 + Z\sqrt{\frac{\sigma_1^2}{n} + \frac{\sigma_2^2}{n}}$$

$$A_H = .5 + (-.21)\sqrt{\frac{20.8}{n}}$$

Setting the two equations for A_H equal to each other we get:

$$0 + 1.645\sqrt{\frac{20.8}{n}} = .5 + (-.21)\sqrt{\frac{20.8}{n}}$$

Solving for n:

$$(1.645)\sqrt{\frac{20.8}{n}} = .5 - .21\sqrt{\frac{20.8}{n}}$$

$$8.46 = .5\sqrt{n}$$

$$\left(\frac{8.46}{.5}\right)^2 = n$$

$$n = 286$$

By allowing alpha to increase from .04 to .10, the required sample size is reduced from 429 to 286. This is due to a relaxing of the alpha constraint which produces a bigger rejection region.

13.19 H_o: $\mu_x \geq 30$

H_A: $\mu_x < 30$

$\alpha = .05$

One tailed test:

The first step is to solve for $\bar{x}$ and S_x:

$$\bar{x} = \frac{\sum_{i=1}^{n} x_i}{n} = \frac{298}{10} = 29.8$$

$$S_x = \frac{\sum x_i^2 - (\sum x_i)^2/n}{n-1} = \frac{13558 - \frac{(298)^2}{10}}{9} = 22.79$$

Next we test the null hypothesis using the t-distribution.

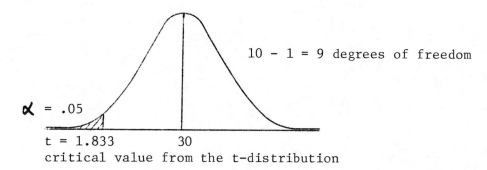

10 - 1 = 9 degrees of freedom

$\alpha = .05$

t = 1.833   30

critical value from the t-distribution

Test:

$$t = \frac{\bar{x} - \mu_x}{\frac{S_x}{\sqrt{n}}}$$

$$t = \frac{29.8 - 30.0}{\frac{22.79}{\sqrt{10}}}$$

$$t = -.027$$

Since t = -.027 > critical t = -1.833 we must <u>not reject</u> the null hypothesis.

While this could imply that a study is needed to see if a children's park is required, six of the ten individuals sampled were ten years of age or younger. Had the null hypothesis been formed as:

$$H_o: \mu_x \leq 30$$

$$H_A: \mu_x > 30$$

The sample data would support this null hypothesis and no study would be undertaken.

13.21 $H_o: \mu_M = \mu_F$

$H_A: \mu_M \neq \mu_F$

$\alpha = .05$

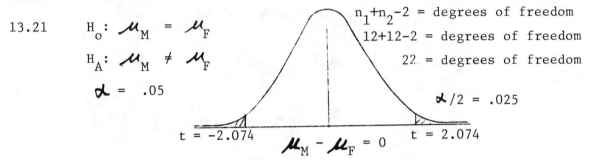

$n_1 + n_2 - 2$ = degrees of freedom
$12 + 12 - 2$ = degrees of freedom
22 = degrees of freedom

$\alpha/2 = .025$

$t = -2.074$ $\mu_M - \mu_F = 0$ $t = 2.074$

First calculate the sample statistics from the raw data.

MALES	FEMALES
$\bar{x}_1 = \dfrac{\Sigma x_i}{n}$	$\bar{x}_2 = \dfrac{\Sigma x_i}{n}$
$\bar{x}_1 = \dfrac{34.01}{12}$	$\bar{x}_2 = \dfrac{36.96}{12}$
$\bar{x}_1 = 2.834$	$\bar{x}_2 = 3.08$
$S_1 = \sqrt{\dfrac{\Sigma x_i^2 - \dfrac{(\Sigma x_i)^2}{n}}{n-1}}$	$S_2 = \sqrt{\dfrac{\Sigma x_i^2 - \dfrac{(\Sigma x_i)^2}{n}}{n-1}}$
$S_1 = \sqrt{\dfrac{101.53 - 96.36}{11}}$	$S_2 = \sqrt{\dfrac{116.43 - 113.84}{11}}$
$S_1 = .683$	$S_2 = .486$

The test:

$$t = \dfrac{\bar{x}_1 - \bar{x}_2 - (\mu_1 - \mu_2)}{S_{pooled}\sqrt{\dfrac{1}{n_1} + \dfrac{1}{n_2}}}$$

Where:

$$S_{pooled} = \sqrt{\dfrac{S_1^2(n_1 - 1) - S_2^2(n_2 - 1)}{n_1 + n_2 - 2}}$$

$$S_{pooled} = \sqrt{\frac{(.683)^2 (11) + (.486)^2 (11)}{22}}$$

$$S_{pooled} = .592$$

Thus:
$$t = \frac{(2.834 - 3.08) - 0}{.592 \sqrt{\frac{1}{12} + \frac{1}{12}}}$$

$$t = -1.01$$

Since t = -1.01 <u>does</u> <u>not</u> fall in the rejection region, the registrar should <u>not</u> conclude that there is a significant difference in GPA's for males and females at Inept Tech.

However, to test this hypothesis using the t-test, the following assumptions are required:

(a) The two populations from which the samples are selected are approximately normally distributed.

(b) Population variances are equal.

13.23 Changing standard deviations to variances we test the hypothesis as follows:

$H_o: \sigma^2 \leq 9$

$H_A: \sigma^2 > 9$

$\alpha = .05$

$n = 30$

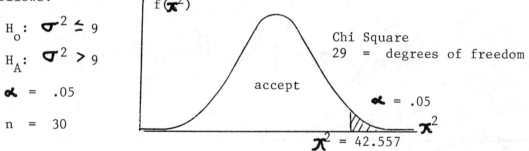

Chi Square
29 = degrees of freedom

$\alpha = .05$

$\chi^2 = 42.557$

The appropriate test statistic

$$\chi^2_{cal} = \frac{(n-1) S_x^2}{\sigma_o^2}$$

The decision rule is:

Accept if $\chi^2_{cal} < \chi^2 = 42.557$

Otherwise reject H_o

$$\chi^2 = \frac{(29)\, 17.64}{9}$$

$$\chi^2_{cal} = 56.84$$

Since $\chi^2 = 56.84 >$ critical value $= 42.557$, the null hypothesis should be <u>rejected</u>.

The manager should conclude that the species of trees planted in this particular field exhibit more than the acceptable amount of variation.

13.25 This is a hypothesis testing problem with the parameter of interest, p, the population proportion. The appropriate null and alternative hypotheses are:

$$H_o: p \geq .20$$

$$H_A: p < .20$$

$$\alpha = .10$$

$$\hat{p} = \frac{18}{100} = .18$$

Our approach in testing this hypothesis is as follows:

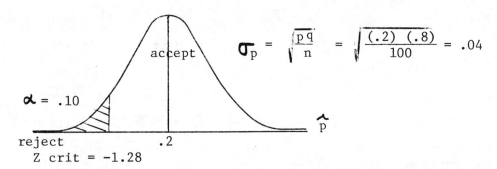

$$\sigma_p = \sqrt{\frac{pq}{n}} = \sqrt{\frac{(.2)(.8)}{100}} = .04$$

$\alpha = .10$

reject
Z crit $= -1.28$

First determine how many standard deivations $\hat{p}$ is from the hypothesized value of p.

$$Z = \frac{\hat{p} - p}{\sigma_p}$$

$$Z = \frac{18 - .20}{.04}$$

$$Z = \frac{-.02}{.04}$$

$$Z = -.5$$

Since Z = -.5 > -1.28, the sample evidence leads us to accept H_o. Thus we <u>cannot</u> support the registrar's claim.

13.27   We need to test a hypothesis where the parameter of interest is p, the population proportion. The appropriate null and alternative hypotheses are:

H_o: p ≥ .40

H_A: p < .40

α = .01

Where $\hat{p} = \frac{18}{49} = .367$

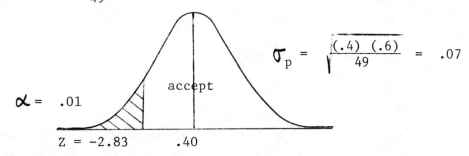

$\sigma_p = \sqrt{\frac{(.4)(.6)}{49}} = .07$

α = .01

Z = -2.83   .40

We solve first for A as:

A = μ_p - Z σ_p

   = .40 + (-2.33) (.07)

   = .237

Since $\hat{p}$ = .367   we cannot reject the null hypothesis.

13.29   We are asked to compute Beta under the assumption that the true value of p = .38. The null and alternative hypotheses are:

H_o: p ≥ .40

H_A: p < .40

α = .01

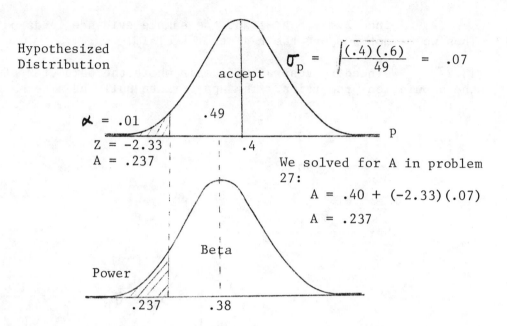

To find Beta:

$$Z = \frac{.237 - .38}{\sqrt{\frac{(.38)(.62)}{49}}}$$

Note: we use .38 in determining the standard deviation because .38 is assumed to be the true population proportion.

$$Z = -2.06$$

From the normal distribution table, the area corresponding to $Z = -2.06$ is .4803. Therefore,

Beta = .5000 + .4803

Beta = .9803

13.31 We must determine the sample size required to satisfy the following constraints:

(a) maximum 5 percent chance of a Type I error
(b) no more than a 10 percent chance of a Type II error when $\mu = 390$ hours
(c) no more than a 14 percent chance of a Type II error when $\mu = 395$ hours

The hypothesis being tested is:

$H_o: \mu_x \geq 400$ hours

$H_A: \mu_x < 400$ hours

156

The errors satisfying the three conditions are shown below:

(a) $A = (-1.645) \dfrac{20}{\sqrt{n}} + 400$

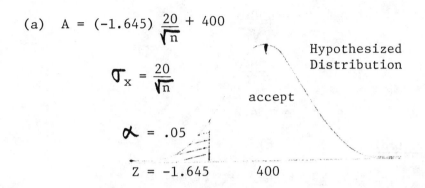

(b)

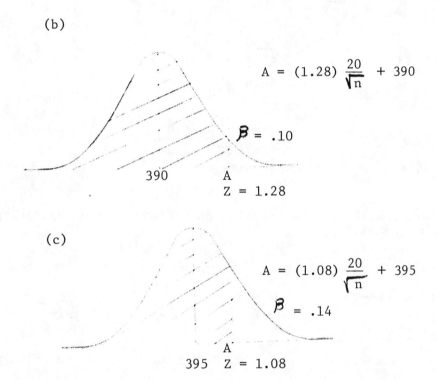

$A = (1.28) \dfrac{20}{\sqrt{n}} + 390$

$\beta = .10$

(c)

$A = (1.08) \dfrac{20}{\sqrt{n}} + 395$

$\beta = .14$

To determine the required sample size, we first analyse constraint (a) and (b), then (a) and (c). We will then select the largest of the two required sample sizes since that would clearly satisfy all constraints.

We proceed as follows:

<u>Constraint (a) and (b):</u>

Set two equations for A equal to each other and solve for n.

$$(-1.645)\frac{20}{\sqrt{n}} + 400 = (1.28)\frac{20}{\sqrt{n}} + 390$$

$$10 = \frac{58.5}{\sqrt{n}}$$

$$n = \frac{58.5}{10}$$

$$n = 34.2 = 35$$

Constraints (a) and (c)

$$(-1.645)\frac{20}{\sqrt{n}} + 400 = (1.08)\frac{20}{\sqrt{n}} + 395$$

$$5 = \frac{54.5}{\sqrt{n}}$$

$$n = \frac{54.5}{5}$$

$$n = 118.81 = \underline{119}$$

Thus, the required sample size to satisfy all constraints is $\underline{119}$ drills.

13.33 We are testing about a population variance. The appropriate null and alternative hypotheses are:

H_o: $\sigma^2 \leq 400$

H_A: $\sigma^2 > 400$

Decision Rule:

Take a sample of $n = 20$, determine S_x^2 and the calculated χ^2:

If $\chi^2 \leq \chi^2_{crit}$, accept H_o

$\chi^2 > \chi^2_{crit}$, reject H_o

The α value and χ^2_{crit} are shown below:

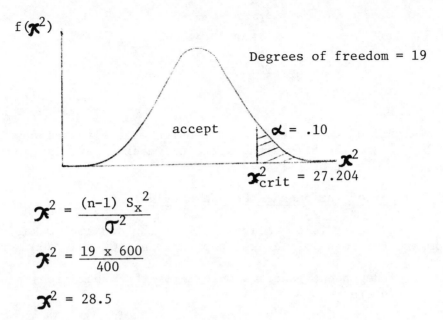

$$\chi^2 = \frac{(n-1) S_x^2}{\sigma^2}$$

$$\chi^2 = \frac{19 \times 600}{400}$$

$$\chi^2 = 28.5$$

Since $\chi^2 > \chi^2_{crit}$, we will reject the claim. Note, our decision is quite sensitive to the level of alpha because our calculated χ^2 is close to the χ^2 critical value. A slight decrease in α would lead us to accept it rather than reject it.

CHAPTER 14

ANALYSIS OF VARIANCE

Analysis of variance is an extension of the techniques from Chapters 12 and 13 which allows the decision maker to examine more than two population means simultaneously.

14.1 Analysis of Variance -- One-Way Design

One-way analysis of variance allows the decision maker to statistically compare the means of several populations. One-way ANOVA requires

1. independent random samples from the populations of interest

2. that all the populations are normal

3. that all the population variances are equal.

Then ANOVA allows the decision maker to test the equality of K means and be certain that the chance of committing a Type I Error is at most alpha.

The appropriate null and alternate hypotheses in one-way ANOVA are

$$H_o: \mu_1 = \mu_2 = \cdots = \mu_K$$

$H_a:$ not all K means are equal

If the three requirements hold, then any differences between the K populations must be differences in means.

We can construct an ANOVA table which summarizes the sample information and leads us to a test statistic. The test statistic is an F-ratio and must be compared to a decision rule based on an F-distribution with appropriate degrees of freedom.

A sample ANOVA table is presented below.

ANOVA Table

Source of Variation	SS	DF	MS	F-Ratio
Between Samples	SSB	K-1	MSB	F
Within Samples	SSW	N-K	MSW	
Total Variation	TSS	N-1		

In this table:

SSB = sum of squares between
SSW = sum of squares within
TSS = total sum of squares
K = number of samples (populations)
N = total number of observations in all samples combined
MSB = $\dfrac{SSB}{K-1}$

MSW = $\dfrac{SSW}{N-K}$

and

F = $\dfrac{MSB}{MSW}$

The decision rule for any one-way analysis of variance utilizes an F-distribution with D_1 = K-1 degrees of freedom (DF) in the numerator and D_2 = N-K degrees of freedom in the denominator. All ANOVA decision rules should dictate rejection of the null hypothesis only for large values of the calculated F statistic. The decision rules are all one-sided in the right hand tail and we should reject H_O if the calculated F exceeds the tabled F.

14.2   Tukey's Method of Multiple Comparison

Tukey's method of multiple comparison can be used to decide which population means differ after the null hypothesis has been rejected in an ANOVA test. All the assumptions necessary to employ this procedure are the same as those for ANOVA except that the sample sizes must all be equal.

Tukey's procedure focuses on the absolute difference between any pair of sample means to decide whether that pair of population means differ. This procedure is conservative since it does not compound the probability of one or more Type I Errors as multiple t-tests would do.

Since Tukey's procedure is conservative it is only applied after the null hypothesis is rejected in the ANOVA test. In addition, the ANOVA procedure could indicate that at least one mean differs but Tukey's procedure might not be able to detect which one.

14.3   Scheffe's Method of Multiple Comparison

Scheffe's procedure, like Tukey's, should only be used after the ANOVA procedure has dictated rejection of the hypothesis of equal means. This procedure is also conservative and does not compound the probability of committing one or more Type I Errors. Scheffe's procedure, unlike Tukey's, does not require that the sample sizes be equal. Because it applies when the sample sizes differ, a new critical value must be computed for each pairwise comparison of sample means.

14.3   Other Analysis of Variance Designs

Many ANOVA designs besides the one-way procedure can be used to compare multiple population means. Each design requires slightly different data collection procedures but can be more efficient then the one-way procedure. See the list of references in the text for sources of additional information regarding other ANOVA designs.

****
*   *
*   SOLUTIONS   *
*   *
****

14.1   (a) Within group variation refers to the fact that the observations taken from a particular population may not all have the same values. The extent to which these values from the particular group vary constitutes the within group variation. In ANOVA we determine this quantity by calculating the sum of squared differences of each value from the respective group mean.

(b) The between group variation refers to the fact that the sample means from the various groups or populations are not all equal. The more variation there is between these sample means the greater is the likelihood that the populations involved have different means.

(c) The total sum of squares represents a measure of the variation in all of the data from all of the groups or populations involved. We calculate this value by finding the sum of the squared differences between each observation and the mean of all observations.

(d) Degrees of freedom is a very difficult concept. One way of thinking of degrees of freedom is that degrees of freedom equal the number of free choices available in arriving at a value of a statistic. Another explanation is that you start with the number of observations available and lose one degree of freedom for each parameter that has to be estimated less one more degree of freedom.

14.3   (a) We first calculate:

$$SSB = \sum_{i=1}^{K} n_i (\bar{x}_i - \bar{\bar{x}})^2$$

Where:
$\bar{x}_1 = 13.60$

$\bar{x}_2 = 12.96$

$\bar{x}_3 = 14.65$

$\bar{\bar{x}} = 13.79$

SSB = $(5)(13.60 - 13.79)^2 + (3)(12.96 - 13.79)^2 + (4)(14.65 - 13.79)^2$

SSB = 5.2056

And:

$$SST = \sum_{i=1}^{K} \sum_{j=1}^{n_j} (x_{ij} - \bar{\bar{x}})^2$$

$$= (13.3 - 13.79)^2 + (14.3 - 13.79)^2 \ldots (14.5 - 13.79)^2$$

$$= 8.3892$$

Then:

SSW = SST - SSB

SSW = 8.3892 - 5.2056

SSW = 3.1836

In table form we get:

ANOVA

Source of Variation	SS	DF	MS	F-Ratio
Between populations	SSB = 5.2056	K - 1 = 2	2.6028	7.528
Within populations	SSW = 3.1836	N - K = 9	.3537	
Total	SST = 8.3892	N - 1 = 11		

H_o: $\mu_1 = \mu_2 = \mu_3$

H_A: not all means are equal

α = .01

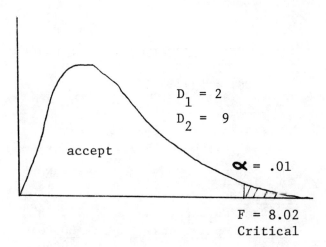

$D_1 = 2$
$D_2 = 9$
accept
α = .01
F = 8.02
Critical

Since F = 7.528 < critical value from F distribution = 8.02 we should <u>not</u> reject the null hypothesis.

(b) Based upon the hypothesis test, the sample evidence does <u>not</u> indicate a significant difference between population means. Therefore, we must assume, based upon this data, that all three types of cars have the same average operating costs.

(c) The primary advantage of equal sample sizes occurs when the null hypothesis is rejected. We can then use Tukey's method of comparison for determining which means are not equal. Tukey's method is preferred over Scheffe's (which allows unequal sample sizes) because it is capable of detecting smaller differences in population means for a given sample size.

14.5 (a) $H_o: \mu_1 = \mu_2 = \mu_3$

$H_A:$ not all means are equal

$\alpha = .05$

To develop the appropriate ANOVA Table we must first find:

$$SSB = \sum_{i=1}^{K} n_i (\bar{x}_i - \bar{\bar{x}})^2$$

Where:

$\bar{x}_1 = 1922 \quad \bar{x}_2 = 2188 \quad \bar{x}_3 = 1764 \quad \bar{\bar{x}} = 1958$

$SSB = 5(1922-1958)^2 + 5(2188-1958)^2 + 5(1764-1958)^2$

$SSB = 459160.0$

$$SST = \sum_{i=1}^{K} \sum_{j=1}^{n} (x_{ij} - \bar{\bar{x}})^2$$

$SST = (1950-1958)^2 + (1870-1958)^2 \ldots (1810-1958)^2$

$SST = 539240.0$

Then:

$SSW = 539240.0 - 459160.0 = 80080.0$

ANOVA

Source of Variation	SS	DF	MS	F-Ratio
Between Populations	459160.0	2	229580	34.40
Within Populations	80080.0	12	6673.3	
TOTAL	539240.0	14		

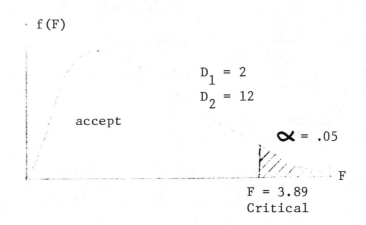

Decision Rule:

If: F-Ratio > critical F with $D_1 = 2$ and $D_2 = 12$ degrees of freedom then reject H_o

Otherwise do not reject.

Since 34.4 > 3.89 reject H_o and conclude there is a difference between nylon rope types.

14.7 Probably the most appropriate design would be a two-factor factorial design as follows.

		Display				
		A	B	C	D	E
S T O R E	Large	x_{111} x_{112} . . x_{11n}	x_{121} x_{122} . . x_{12n}			x_{151} . . x_{15n}
S I Z E	Medium	. . . .				
	Small	x_{311} . . x_{31n}				x_{351} . . x_{35n}

Where:

x_{ijk} = Kth measurement for display j for store size i.

To have an equal sample size in each box would require a larger sample size than the 25 in the original problem. The manufacturer will have to decide whether the added cost of sampling will be off-set by the added information obtained.

14.9   This problem requires us to utilize Scheffe's method of comparison because the sample sizes are not equal.

We must calculate the S-Range for each pair of populations. If the absolute value of the difference between any pair of sample means exceeds the S-Range, a significant difference in means will be determined to exist.

$$\text{S-Range} = S\hat{\sigma}$$

Where:

$$S = \sqrt{(K-1)(F_{(1-\alpha)\, D_1,\, D_2})}$$

$$S = \sqrt{(2)(4.26)}$$

$$S = 2.919$$

And:

$$\hat{\sigma} = \sqrt{\left(\frac{1}{n_1} + \frac{1}{n_2}\right) MSW}$$

If we compare population 1 and population 2, $n_1 = 5$ and $n_2 = 3$.

Then:

$$\hat{\sigma} = \sqrt{\left(\frac{1}{5} + \frac{1}{3}\right).3537}$$

$$\hat{\sigma} = .4343$$

Then:

$$\text{S-Range} = (2.919)(.443)$$

$$= 1.267$$

And:

$$|\bar{x}_1 - \bar{x}_2| = |13.6 - 12.96| = .64$$

Since .64 < 1.28 we <u>cannot</u> conclude (at α = .05 level) that populations 1 and 2 have means which differ.

Examining the other possible pairwise comparisons we get:

$$|\bar{x}_1 - \bar{x}_3| = |13.6 - 14.65| = 1.05$$

$$\text{S-Range} = 2.919 \sqrt{(\tfrac{1}{5} + \tfrac{1}{4}) \; .3537} = 1.164$$

Since $1.05 < 1.164$ we cannot conclude that there is a difference in the mean operating costs for car brand 1 and brand 3.

For:
$$|\bar{x}_2 - \bar{x}_3| = |12.96 - 14.65| = 1.69$$

$$\text{S. Range} = 2.919 \sqrt{(\tfrac{1}{3} + \tfrac{1}{4}) \; .3537} = 1.325$$

Since $1.69 > 1.325$ we conclude that there <u>is</u> a significant difference between the average operating costs for cars 2 and 3.

14.11   In problem 14.5, the analysis of variance F test led to rejection of the null hypothesis that all population means were equal. Now we are interested in determining which type of nylon rope should be purchased.

<u>Tukey's Method of Comparison</u>

For any absolute difference between a pair of sample means which exceeds $T\sqrt{MSW}$, conclude the respective population means are unequal.

$$T = \frac{1}{\sqrt{n}} q\,(k, N-K) = \frac{1}{\sqrt{5}} q\,(3, 12)(\alpha = .05) = \frac{1}{\sqrt{5}}(3.77)$$

$$T\sqrt{MSW} = \frac{1}{\sqrt{5}}(3.77)\sqrt{6,673.3}$$

$$= 137.7$$

Now we look at all pairwise comparisons:

$$|\bar{x}_2 - \bar{x}_3| = |2188 - 1764| = 424 > 137.7 \quad \mu_2 \neq \mu_3$$

$$|\bar{x}_2 - \bar{x}_1| = |2188 - 1922| = 266 > 137.7 \quad \mu_2 \neq \mu_1$$

$$|\bar{x}_1 - \bar{x}_3| = |1922 - 1764| = 158 > 137.7 \quad \mu_1 \neq \mu_3$$

Thus we conclude that all the population means are distinct and since $\bar{x}_2 > \bar{x}_1 > \bar{x}_3$ the logical extension is that $\mu_2 > \mu_1 > \mu_3$.

Therefore based on these data, they should purchase brand II.

Notice that if $\alpha = .01$ the ANOVA would still lead to rejection since CV = 6.93 < 34.4. However the application of Tukey's procedure yields $T\sqrt{MSW} = 184.1$ and we would conclude $\mu_2 \neq \mu_3$, $\mu_2 \neq \mu_1$, but μ_1 may equal μ_3. The purchase recommendation would still be, buy brand II.

14.13   We are given: TSS = 2900
                        SSW = 700

Thus we can compute: SSB = SST - SSW

                           = 2900 - 700

                           = 2200

The appropriate null and alternative hypotheses are:

H_o: $\mu_1 = \mu_2 = \mu_3$

H_A: not all means are equal

$\alpha = .05$

We now set up the analysis of variance table:

ANOVA

Source of Variation	SS	DF	MS	F-Ratio
Between	2200	2	1100	108.42
Within	700	69	10.14	
TOTAL	2900	71		

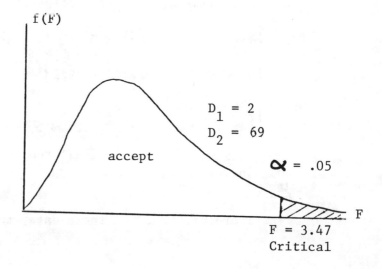

$D_1 = 2$
$D_2 = 69$
$\alpha = .05$
F = 3.47 Critical

Decision Rule:

If the calculated F Ratio > $F_{critical}$ = 3.47 we should reject H_o. Otherwise do not reject H_o.

Since 108.42 > 3.47 we should clearly reject H_o and conclude that the average age is not the same for viewers of the three TV Stations in Bextfort, Washington.

14.15 While the S-Range can be developed, we would stipulate that the Tukey procedure and therefore the T-Range is more appropriate because of equal sample sizes.

The S-Range is developed as follows:

$$\text{S-Range} = S\hat{\sigma}$$

Where:

$$S = \sqrt{(K-1) F_{critical}}$$

$$\hat{\sigma} = \sqrt{\left(\frac{1}{n_1} + \frac{1}{n_2}\right) MSW}$$

Thus:

$$S = \sqrt{(2)\, 3.47}$$

$$S = 2.63$$

And:

$$\hat{\sigma} = \sqrt{\left(\frac{1}{24} + \frac{1}{24}\right)(10.14)}$$

$$\hat{\sigma} = .919$$

Thus:

$$\text{S-Range} = (2.63)(.919)$$

$$\text{S-Range} = 2.416$$

Thus, for any pair of sample means whose absolute difference exceeds 2.416, we would conclude that their respective population means differ.

14.17   T-Range = 1.826

S-Range = 2.416

We see that the T-Range < S-Range. Therefore, the T-Range will be more sensitive to small differences in population means and is thus a more powerful procedure to use when pairwise comparisons are made and the sample sizes are equal.

14.19 $H_o: \mu_1 - \mu_2 = 0$

$H_A: \mu_1 - \mu_2 \neq 0$

$\alpha = .05$

The appropriate test statistic is:

$$t = \frac{\bar{x}_1 - \bar{x}_2 - 0}{S_p \sqrt{\frac{1}{n_1} + \frac{1}{n_2}}}$$

Where:

$$S_p = \sqrt{\frac{S_1^2 (n_1 - 1) + S_2^2 (n_2 - 1)}{n_1 + n_2 - 2}}$$

First we must compute $\bar{x}_1$, $\bar{x}_2$, S_1^2 and S_2^2 from the raw data.

$$\bar{x}_1 = \frac{\sum x_i}{n_1} = \frac{151}{9} = 16.77$$

$$\bar{x}_2 = \frac{\sum x_i}{n_2} = \frac{113}{9} = 12.55$$

$$S_1^2 = \frac{\sum_{i=1}^{n_1} x_i^2 - \frac{(\sum x)^2}{n_1}}{n_1 - 1} = \frac{3385 - \frac{(151)^2}{9}}{8} = 106.44$$

$$S_2^2 = \frac{\sum_{i=1}^{n_2} x_i^2 - \frac{(\sum x)^2}{n_2}}{n_2 - 1} = \frac{1515 - \frac{(113)^2}{9}}{8} = 12.02$$

Thus the t-test is:

$$t = \frac{(16.77 - 12.55) - 0}{\sqrt{\frac{(106.44)(8) + (12.02)(8)}{16}} \sqrt{\frac{1}{9} + \frac{1}{9}}}$$

$$t = \frac{4.22}{7.696 \ (.4714)}$$

$$t = 1.16$$

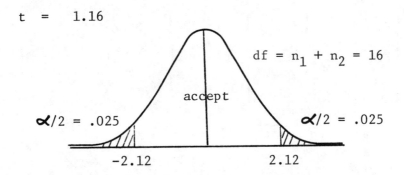

Decision Rule:

If calculated $t > t_{critical} = 2.12$ or if calculated $t < t_{critical} = -2.12$ Reject H_0. Otherwise do not reject.

Therefore since $t = 1.16 < 2.12$ we should <u>not</u> reject the null hypothesis. Based upon these data we cannot conclude a difference between the average numbers in the two lift lines.

14.21 It is clear that both the t test and the analysis of variance F test lead us to the same conclusions. That is, we should <u>not</u> reject the null hypothesis of equal population means. However, further examination reveals that the calculated F value of 1.35 is equal to the calculate t squared.

That is:

$$(1.16)^2 = 1.35$$

This relationship:

$$F = t^2$$

will always hold for the two sample cases.

171

14.23   Since the sample sizes are equal for all groups, we should use Tukey's Method of Multiple Comparison. Tukey's method involves comparing the difference between pairs of sample means to a T-range value.

The pairwise comparisons are called contrasts.

$$|\bar{x}_{Leach} - \bar{x}_A| = |111 - 126| = 15$$

$$|\bar{x}_{Leach} - \bar{x}_B| = |111 - 100| = 11$$

$$|\bar{x}_{Leach} - \bar{x}_C| = |111 - 105| = 6$$

$$|\bar{x}_A - \bar{x}_B| = |126 - 100| = 26$$

$$|\bar{x}_A - \bar{x}_C| = |126 - 105| = 21$$

$$|\bar{x}_B - \bar{x}_C| = |100 - 105| = 5$$

T-Range = $T\sqrt{MSW}$

Where:

$$T = \frac{1}{\sqrt{n}} \, q_{(1-\alpha)} \text{ with } D_1 = K = 4$$

$$D_2 = N - K = 76$$

Common sample size n = 20

From the Studentized Range Table, Appendix 1-G, the q value is approximately 3.72.

Then:

$$T = \frac{1}{\sqrt{20}} (3.72)$$

$$T = .832$$

Therefore:

T-Range = $.832\sqrt{157.89}$

T-Range = 10.45

Looking at the pairwise comparisons again, if any absolute difference in sample mean exceeds 10.45 we can conclude a difference exists at the .05 alpha level.

	(Contrast)				(Significant)
$\lvert \bar{x}_{leach} - \bar{x}_A \rvert$	=	$\lvert 111 - 126 \rvert$	=	15	Yes
$\lvert \bar{x}_{leach} - \bar{x}_B \rvert$	=	$\lvert 111 - 100 \rvert$	=	11	Yes
$\lvert \bar{x}_{leach} - \bar{x}_C \rvert$	=	$\lvert 111 - 105 \rvert$	=	6	No
$\lvert \bar{x}_A - \bar{x}_B \rvert$	=	$\lvert 126 - 100 \rvert$	=	26	Yes
$\lvert \bar{x}_A - \bar{x}_C \rvert$	=	$\lvert 126 - 105 \rvert$	=	21	Yes
$\lvert \bar{x}_B - \bar{x}_C \rvert$		$\lvert 100 - 105 \rvert$	=	5	No

We infer from these results that A > (leach or B or C).

Therefore, given the criteria listed, the sample results indicate that Savovy should switch to Brand A tires for their bicycles.

14.25 Since the sample sizes are different we must use Scheffe's Method of Multiple Comparisons. Scheffe's method compares the difference between pairs of sample means to a S-Range Value. We will compute a different S-Range for each pairwise comparison.

The pairwise comparisons are:
(Contrasts)

(1) $\lvert \bar{x}_{CAN} - \bar{x}_{AUST} \rvert$ = $\lvert 3111 - 2005 \rvert$ = 1106
(2) $\lvert \bar{x}_{CAN} - \bar{x}_{GERM} \rvert$ = $\lvert 3111 - 3511 \rvert$ = 400
(3) $\lvert \bar{x}_{AUST} - \bar{x}_{GERM} \rvert$ = $\lvert 2005 - 3511 \rvert$ = 1506

For each comparison we compute:

S-Range = $S\hat{\sigma}$

Where:

$$S = \sqrt{(K-1)F_{critical}}$$

$$S = \sqrt{(3-1)3.55}$$

$$S = 2.66$$

And:

$\hat{\sigma}$ will vary since

$$\hat{\sigma} = \sqrt{(\frac{1}{n_i} + \frac{1}{n}) \text{MSW}}$$

For the Canadians and Austrians comparison

$$\hat{\sigma} = \sqrt{(\frac{1}{6} + \frac{1}{8}) \, 509114.67}$$

$$\hat{\sigma} = 385.34$$

Then:

$$\text{S-Range} = 2.66 \times 385.34$$

$$= 1025.02$$

For the Canadians and Germans comparison:

$$\hat{\sigma} = \sqrt{(\frac{1}{6} + \frac{1}{7}) \, 509114.67}$$

$$\hat{\sigma} = 396.97$$

$$\text{S-Range} = 2.66 \times 396.97$$

$$= 1055.93$$

For the Austrians and Germans comparison

$$\hat{\sigma} = \sqrt{(\frac{1}{8} + \frac{1}{7}) \, 509114.67}$$

$$= 369.28$$

$$\text{S-Range} = 2.66 \times 369.28$$

$$= 982.29$$

Comparing each contrast with its S-Range:

Pair	Constrast	S-Range	Significant
Canadians and Austrians	1106	1025.02	Yes
Canadians and Germans	400	1055.93	No
Austrians and Germans	1506	982.29	Yes

Based upon these results, we infer that the Austrians receive lower average earnings from private lessons than either the Canadians or the Germans.

CHAPTER 15

HYPOTHESIS TESTING USING NONPARAMETRIC STATISTICS

The previous chapters relating to statistical inference have introduced you to a variety of statistical tests. Each of these tests required that the available data to be at least interval scaled. Further, certain of the tests such as the t test and the F test for analysis of variance require rather restrictive assumptions about the shape of the population distribution. Specifically, these tests are based upon the assumption that the population is normally distributed.

These assumptions and data measurement requirements often prove too restrictive for many decision-making situations. In many instances, the level of data measurement is either nominal or ordinal and in other cases the normality assumption may clearly be violated. Thus, decision makers have a real need for statistical techniques which allow flexibility of data measuremend and distribution assumptions. Nonparametric statistics are a group of statistical techniques which have been developed to do exactly this.

In chapter 15, several of the most commonly used nonparametric tests are introduced and business examples are presented.

15-1 Chi-Square Goodness of Fit Test:

Many situations exist in which the decision maker has reason to believe a distribution or specific pattern may exist and he is interested in determining whether sample data support this belief. For example, the manager of a franchise chicken establishment hires personnel and orders food on the basis that 60% of his weekly business is evenly spread between Fridays and Saturdays. The remaining 40% is spread between Monday through Thursday. He is closed Sunday.

The Chi-Square Goodness of Fit Test is the appropriate statistical procedure to test the manager's assumption about customer distribution through the week. Over a period of time, the manager would keep track of the number of customers arriving on each of the six days. The Chi-Square test then compares the number that actually arrived on each day with the number that would be expected under the manager's assumptions. The Chi-Square test statistic is:

$$\chi^2 = \frac{\Sigma(f_o - f_e)^2}{f_e}$$

where:

$$f_o = \text{observed frequency}$$

$$f_e = \text{expected frequency}$$

Note, if the observed and expected frequencies match closely, χ^2 value will be small. If the observed and expected frequencies differ greatly, a large χ^2 will result. Thus the manager would <u>reject</u> his assumption about customer distribution if the χ^2 value gets too large. To determine if the χ^2 is too large, we compare if to a table value from the Chi-Square distribution for the specified alpha level and degrees of freedom equal to k-1, where k equals the number of categories. In the chicken establishment example, the number of categories is 6 corresponding to six days so the degree of freedom is 6-1 = 5.

15-2 Chi-Square Goodness of Fit Limitations:

There are certain limitations or restrictions connected with the Chi-Square Goodness of Fit test:

1. When the degrees of freedom equal 1.0, the expected frequencies in each category should be at least 5.

2. In all other cases (df > 1) at least 80% of the categories should have expected cell frequencies greater than 5 and all categories should have expected frequencies greater than 1.0.

Generally, these limitations can be overcome by increasing the sample size and/or combining categories to increase the expected frequencies.

15-3 Mann-Whitney U Test:

A nonparametric statistical test whose parametric counterpart is the two sample t-test is called the Mann-Whitney U Test. This test applies when two samples are assumed independent and the data are at least ordinal. This procedure is used to test whether the data come from populations with the same means and standard deviations.

The Mann-Whitney U test works by first converting the data to rankings. The U values are computed for both samples as follows:

$$U_A = (n_1)(n_2) + \frac{(n_1)(n_1 + 1)}{2} - \Sigma R_1$$

$$U_B = (n_1)(n_2) + \frac{(n_2)(n_2 + 1)}{2} - \Sigma R_2$$

Then for a <u>two tailed test</u>, we pick either U_A or U_B and refer to this as U_{test}. If the test is <u>one tailed</u>, set U_{test} equal to whichever of U_A or U_B is most likely to fall in the rejection region.

Providing n_1 and n_2 are both at least 10, the Mann-Whitney U Test statistic is:

$$Z = \frac{U_{test} - \frac{(n_1)(n_2)}{2}}{\sqrt{\frac{(n_1)(n_2)(n_1 + n_2 + 1)}{12}}}$$

Note, this is the standard normal statistic and the hypothesis test is carried out by comparing the Z value with table Z values from the normal table in the text.

The logic of the Mann-Whitney U test is based around the idea that the weighted sum of the rankings will be nearly equal if the samples actually do come from the populations with the same distribution (i.e., means and standard deviations are equal).

The text does not discuss the methods which must be employed to apply the Mann-Whitney U test when the sample sizes are small nor does it discuss how to make the necessary correction when a large number of ties exist in the rankings. Both of these topics are addressed in texts devoted solely to nonparametric statistics such as those referenced in the text.

15-4 Kolmogorov-Smirnov Two Sample Test--Large Samples:

A comparable test to the Mann-Whitney U test is the Kolmogorov-Smirnov Two Sample Test. This test can be used if you wish to test whether two independent samples come from populations with the same distribution and the data you have are at least ordinal.

The <u>one tailed</u> test statistic when both samples are larger than 40 is:

$$\chi^2 = 4 D^2 \left[\frac{(n_1)(n_2)}{n_1 + n_2}\right]$$

where:

D - maximum difference between cummulative relative frequencies

By examining the above formula we see that the χ^2 statistic becomes large as D becomes large. Thus if the two samples differ, the D value will be large leading to a large χ^2. The hypothesis that the two population distributions are the same will be rejected if the samples differ extensively since we reject H_o if χ^2 is greater than a chi-square table value with <u>two</u> degrees of freedom.

If the test is <u>two tailed</u>, we simply compute D, the maximum absolute difference in cummulative relative frequency and compare this value to the table of D value in Appendix H of the text. If the computed D exceeds the table value, we conclude the two populations have different distributions.

15-5 <u>Contingency Analysis</u>:

Contingency analysis is actually an extension of Chi-Square Goodness of Fit testing but rather than one variable, we are interested in two variables. Specifically, <u>contingency analysis can be used to test the independence of two variables</u>. For example, suppose during a political poll, people were asked what their party was: Republican, Democrat, Libertarian, Independent, or Other. Also, the respondent's sex was determined as male or female. The following contingency table reflects the results of a sample of 1,000 people.

	Republican	Democrat	Libertarian	Independent	Other	Total
Male	200	300	150	20	30	700
Female	100	100	50	50	0	300
Total	300	400	200	70	30	1000

Now, the question is, do the data indicate that party preference is independent of the respondent's sex?

The contingency analysis approach is to determine the expected frequency for each cell and then see how well the observed frequencies fit the expected frequencies. This is analogous to the Chi-Square Goodness of Fit test concept.

The test statistic is computed as follows:

$$\chi^2 = \sum_{row} \sum_{col.} \frac{(f_o - f_e)^2}{f_e}$$

where:

f_o = observed frequency

f_e = expected frequency

Thus, if the expected and observed frequencies are closely matched, the χ^2 value will be small and will lead to the conclusion that the two variables are independent. Otherwise if χ^2 is large, we conclude that some form of dependency exists. In any case, we compare the χ^2 with a table Chi-Square value with degrees of freedom equal to (rows - 1)(columns - 1).

The <u>expected cell frequencies</u> are found by:

$$f_{e_{ij}} = \frac{(\Sigma\,row\,i)(\Sigma\,column\,j)}{Total\ Observations}$$

As with the goodness of fit test, there is a limitation on the use of contingency analysis. Specifically, <u>all cells must have expected cell frequencies of at least five</u>. This can be accomplished by having a large sample size or by grouping categories into logical sets.

15-6 <u>Kruskal-Wallis One Way Analysis of Variance</u>:

The Kruskal-Wallis One Way Analysis of Variance represents the non-parametric alternative to the analysis of one way variance procedure presented in Chapter 14. It has the advantage of not requiring that the populations be normal and it can be utilized when the data are at least of ordinal measurement.

The Kruskal-Wallis procedure is based upon rankings within each sample and the test statistic is an H statistic computed as follows:

$$H = \frac{12}{N(N+1)} \sum_{i=1}^{k} \frac{R_i}{n_i} - 3(N+1)$$

where:

N = total of all observations

R_i = sum of the ranks in the ith sample

n_i = sample size of the ith sample

K = number of samples

This H statistic is distributed as Chi-Square statistic with K-1 degrees of freedom if the samples actually do come from populations with equal means. A large H-statistic leads to rejection of the null hypothesis of equal means.

In order for the Kruskal-Wallis one way analysis of variance to apply, we must have <u>independent samples with sizes of at least 5</u>. Also, if we have ties in the ranking, the H statistic must be modified. With the correction for ties, the H statistic is determined by:

$$H = \frac{\frac{12}{N(N+1)} \sum_{i=1}^{k} \frac{R_i^2}{n_i} - 3(N+1)}{1 - \frac{\sum_{i=1}^{g}(t_i^3 - t_i)}{N^3 - N}}$$

where:

t = number of tied cases in each group of tied scores

N = total number of observations

g = number of different groups of ties

R_i = sum of rankings in ith sample

15-7 <u>Conclusions</u>:

The nonparametric techniques presented in Chapter 15 represent only a few of the many such techniques available to the decision maker. All the nonparametric tests share the features of reduced data measurement requirements and relaxed distribution assumptions as compared to the statistical tests presented in earlier chapters. These can be of real value to decision makers with data measurement limitations.

You should remember, however, that in those situations where a parametric test such as the t test could be appropriately employed, the use of a corresponding nonparametric test will require an increased sample size to retain the same power (probability of rejecting a false hypothesis) as the t test.

15-1 A parametric statistical test has certain assumptions associated with it about the level of data measurement and the distribution of the population(s) from which the data are collected. A nonparametric test provides the opportunity to test certain hypotheses without making such restrictive assumptions. Nonparametric tests are often called distribution-free tests because they make no specific assumptions about the populations from which the data are collected.

15-3 The manager should establish the following null hypothesis:

H_o : The use of checkout stands is uniform. (An equal number of people will use each cehckout stand.)

H_a : The use of checkout stands is not uniform.

Additionally, to test this null hypothesis, we must select a desired alpha level. Keeping in mind the costs of making a Type I and a Type II error, we have selected $\alpha = .05$. Other decision makers might well select a different alpha level.

The appropriate test is the Chi-Square Goodness of Fit test in which the test statistic is calculated as follows:

$$\chi^2 = \sum \frac{(f_o - f_e)^2}{f_e}$$

where:

f_o = observed frequency in each group

f_e = expected frequency in each group if H_o is true

	Checkout Stand				
	1	2	3	4	Total
f_o	292	281	240	187	1000
f_e	250	250	250	250	1000

then:

$$\chi^2 = \frac{(292-250)^2}{250} + \frac{(281-250)^2}{250} + \frac{(240-250)^2}{250} + \frac{(187-250)^2}{250}$$

$$\chi^2 = 27.176$$

We compare the calculated Chi-Square statistic to the Chi-Square tabled value with k-1 = 3 degrees of freedom for alpha = .05.

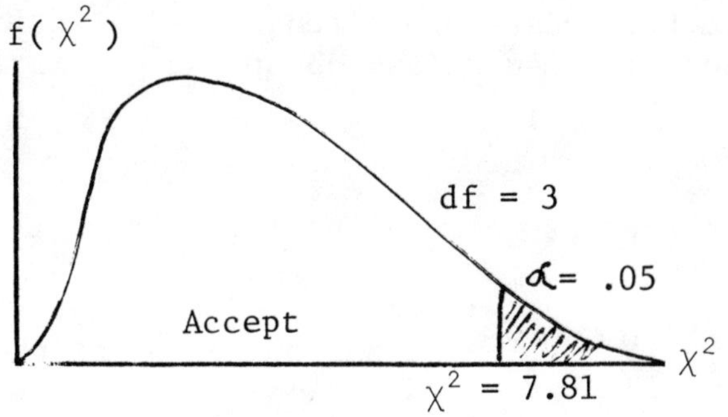

Decision Rule: If calculated Chi-Square > critical value from table, reject H_o, otherwise accept (do not reject).

Therefore, since

$\chi^2 = 27.176 > 7.81$, we conclude that the sample evidence is sufficient to cause us to reject the null hypothesis. We would conclude that all checkout stands do not receive the same volume of customers.

15-5 (a)

H_o : client income is independent of region

H_a : client income is not independent of region

$\alpha = .05$

The appropriate statistical test is the Chi-Square Contingency Test.

		Region				
		Northeast	South	Midwest	West	Total
Income	Less $25,000	$f_o = 57$ $f_e = 75.75$	$f_o = 112$ $f_e = 106.88$	$f_o = 109$ $f_e = 88.89$	$f_o = 71$ $f_e = 77.48$	349
	Greater $25,000	$f_o = 162$ $f_e = 143.25$	$f_o = 197$ $f_e = 202.12$	$f_o = 148$ $f_e = 168.10$	$f_o = 153$ $f_e = 146.5$	660
	Total	219	309	257	224	1009

We find the expected cell frequencies as follows:

$$f_e = \frac{\text{Row total} \times \text{Column total}}{\text{Grand total}}$$

$$f_e = \frac{349 \times 219}{1009} = 75.75$$

$$\vdots$$

$$f_e = \frac{660 \times 224}{1009} = 146.52$$

then:

$$\chi^2 = \sum\sum \frac{(f_o - f_e)^2}{f_e}$$

$$\chi^2 = \frac{(57-75.75)^2}{75.75} + \frac{(112 - 106.88)^2}{106.88} \ldots + \frac{(153-146.52)^2}{146.52}$$

$$\chi^2 = 15.25$$

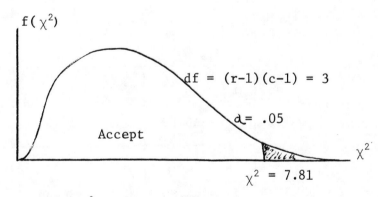

Decision Rule: If: Calculated $\chi^2 > 7.81$ reject the null hypothesis, otherwise do not reject.

Since: The calculated value, $\chi^2 = 15.25 > 7.81$ we <u>reject</u> the null hypothesis. Based upon the sample evidence, we conclude that there is a dependency relationship between client income level and region. One of the regions (probably Midwest) may have been more effective in attracting the clients with incomes under $25,000.

We selected an alpha level equal to .05. Obviously the level selected depends upon the decision maker's willingness to commit a Type I error.

We set this probability fairly low feeling that to reject H_0 when it is true may be more serious than accepting H_O if it is false. If we did commit a Type I error, we would conclude that there was a dependency between client income and region when actually there is no such relationship. Based upon this incorrect decision, we might take unwarranted action to step up the effectiveness in the other regions. This was viewed as more serious than the Type II error.

15-7 (a)

H_o : propensity for retreatment is independent of treatment price

H_a : propensity for retreatment is not independent of treatment price

$\alpha = .10$

Note, we have selected an alpha level = .10 based upon our assessment of the costs involved with making a Type I error. Any other decision maker may have selected a different alpha. Also, the null hypothesis must be set up indicating independence rather than dependence like the accountant claimed since we otherwise would have to specify exactly what the relationship is.

	High	Medium	Low	Total
Less than 1	$f_o = 26$ $f_e = 18.89$	$f_o = 13$ $f_e = 15.38$	$f_o = 16$ $f_e = 20.71$	55
2 - 5	$f_o = 43$ $f_e = 44.66$	$f_o = 35$ $f_e = 36.36$	$f_o = 52$ $f_e = 48.96$	130
not in 5	$f_o = 87$ $f_e = 92.43$	$f_o = 79$ $f_e = 75.24$	$f_o = 103$ $f_e = 101.31$	269
	156	127	171	454

We find the cell expected frequencies as follows:

$$f_e = \frac{\text{row total} \times \text{column total}}{\text{grand total}}$$

than:

$$f_e = \frac{(55 \times 156)}{454} = 18.89$$

$$\vdots$$

$$f_e = \frac{(269 \times 171)}{454} = 101.31$$

then:

$$\chi^2 = \Sigma\Sigma \frac{(f_o - f_e)}{f_e}$$

$$\chi^2 = \frac{(26-18.89)^2}{18.89} + \ldots \ldots \frac{(103-101.31)^2}{101.31}$$

$$\chi^2 = 4.952$$

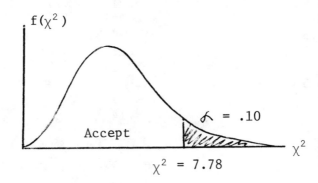

$\chi^2 = 7.78$

Decision Rule: If: Calculated $\chi^2 > 7.78$, reject the null hypothesis, otherwise do not reject H_o.

Since: $\chi^2 = 4.952 < 7.78$, we <u>do not</u> reject the null hypothesis. Based upon the sample information, the accountant <u>cannot</u> conclude that a significant relationship exists between price and re-treatments.

(b) The accountant has ignored such potentially important factors as:

- the type of injury or problem for which the the treatment was originally given
- travel distance to Tsu Chi Chun's office
- the individual's income

15-9

$$H_o : \mu_1 = \mu_2 = \mu_3$$

H_a : not all population means are equal

$\alpha = .10$

Note, we have selected alpha equal to .10. Any other decision maker may have selected a different alpha depending upon his or her willingness to make a Type I error. Also the relationship between alpha and the chances of committing a Type II error should also be considered.

The first step in testing the hypothesis is to convert the actual measurements into their relative rankings (lowest to highest) as follows:

Method 1	Method 2	Method 3
24	17.5	1
32	15	7
31	7	9
33	21	19
10	22	23
26.5	29.5	20
26.5	26.5	11.5
29.5	2	3
26.5	4	5
17.5	7	11.5
	14	13
		16
$\Sigma = 256.5$	$\Sigma = 165.5$	$\Sigma = 139.0$

Number of cases involved in tied rankings:

At 31 rank = 7   number = 3
At 34 rank = 11.5 number = 2
At 39 rank = 11.5 number = 2
At 46 rank = 26.5 number = 4
At 47 rank = 29.5 number = 2

13 observations involved in ties

Since 13 cases are involved in ties, we need to utilize the correction for ties.

The appropriate test statistic is:

$$H = \frac{\frac{12}{N(N+1)} \sum_{i=1}^{g} \frac{R_1^2}{n_1} = 3(N+1)}{1 - \frac{\sum_{i=1}^{K}(t_i^3 - t_1)}{N^3 - N}}$$

Then:

$$H = \frac{\frac{12}{(33)(34)} \left[\frac{(256.5)^2}{10} + \frac{(265.5)^2}{11} + \frac{(139)^2}{12}\right] - 3(34)}{1 - \frac{(3^3-3) + (2^3-2) + (2^3-2) + (4^3-4) + (2^3-2)}{33^3 - 33}}$$

$$H = \frac{12.217}{.9971}$$

$$H = 12.252$$

If the null hypothesis is true, the H statistic is Chi-square distributed with k-1 degrees of freedom.

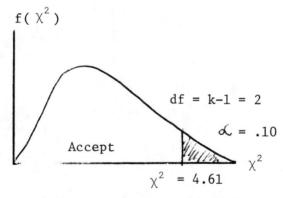

Decision Rule: If: $H >$ critical $\chi^2 = 4.61$ reject H_o, otherwise do not reject.

Since $H = 12.252 > 4.61$, we should <u>reject</u> the null hypothesis based upon this sample information and conclude that <u>not all</u> population means are equal. Thus the pollution control methods would <u>not</u> be considered equal based upon this sample information.

15-11 H_o : There is no difference in the image ratings for Bluedot and Companies A and B

H_a : There is a difference in the image ratings between the three

$\alpha = .10$

We have selected an alpha level equal to .10 reflecting a fairly strong desire to avoid a Type I error. To reject H_O when it is true would be to falsely conclude that a significant difference exists between the three brands of beer. Bluedot might infer that their image is lower than their competitors when in fact it is not.

The appropriate test is the Kruskal-Wallis One Way Analysis of Variance test. The test statistic with the correction factor for ties is:

$$H = \frac{\frac{12}{N(N+1)} \sum_{i=1}^{K} \frac{R_i^2}{n_i} - 3(N+1)}{1 - \frac{\sum(t_i^3 - t_i)}{N^3 - N}}$$

Our first step is to transform the actual data into the appropriate ranks as follows:

Bluedot	A	B
3.5	23	5
10	7	14
12	8.5	15
3.5	20	16
8.5	18.5	21
18.5		6
1.5		11
1.5		13
		22
		17
$\Sigma = 59$	$\Sigma = 77$	$\Sigma = 140$

Ties: 2 tied at score 20
       2 tied at score 40
       2 tied at score 55
       2 tied at score 90

Then:

$$H = \frac{\frac{12}{(23)(24)} \left[\frac{(59)^2}{8} + \frac{(77)^2}{5} + \frac{(140)^2}{10} \right] - 3(24)}{1 - \frac{\left[(2^3-2) + (2^3-2) + (2^3-2) + (2^3-2) \right]}{23^3 - 23}}$$

$$H = 5.857$$

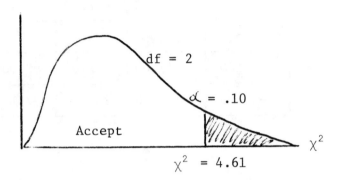

$$\chi^2 = 4.61$$

Decision Rule: If: calculated $H > 4.61$ reject H_o, otherwise do not reject.

Since: $H = 5.857 > 4.61$, the sample evidence leads us to <u>reject</u> the null hypothesis. Bluedot should conclude that there is a difference in image ratings between the three companies. Note, however, that if the alpha level had been .05, the appropriate decision would have been to <u>not</u> reject the null hypothesis, since the critical value of $\chi^2 = 5.99$.

15-13 H_o : Image ratings for Bluedot are at least as good after the federally imposed recall as before.

H_a : Image ratings are lower following the recall than before it was imposed.

$\alpha = .05$

Note, we have selected a fairly small alpha level because in our view a Type I error is considered more serious than a Type II error. Another decision maker might arrive at a different level of alpha depending upon the individual's view of the costs involved with each error. The important thing here is to recognize the cost tradeoffs between the two types of errors which might result.

An appropriate test is the Kolmogorov-Smirnov two sample test, one tailed. The test statistic is:

$$\chi^2 = 4 D^2 \left[\frac{(n_1)(n_2)}{n_1 + n_2} \right]$$

We set up the following table:

Image Rating

	0-19	20-39	40-59	60-79	80-100	Total
Before Freq.	3	28	30	30	9	100
$S_1(x)$	3/100	31/100	61/100	91/100	100/100	
After Freq.	8	36	40	14	2	100
$S_2(x)$	3/100	44/100	84/100	98/100	100/100	
$S_1(x) - S_2(x)$	-5/100	-13/100	-23/100	-7/100	0	

↑
D

Then:

$$\chi^2 = (4)(-23/100)^2 \left[\frac{(100)(100)}{100 + 100} \right]$$

$$\chi^2 = 10.58$$

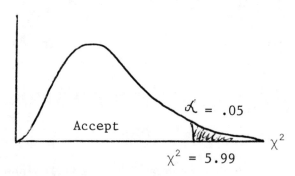

$\chi^2 = 5.99$

Decision Rule: If: Calculated $\chi^2 > 5.99$ reject the null hypothesis, otherwise do not reject.

Since: The $\chi = 10.58 > 5.99$ the sample evidence clearly indicates that the null hypothesis should be <u>rejected</u>. Bluedot should conclude that their image ratings are lower following the recall. If they feel that the recall was not warranted, they might use this information in an attempt to recover damages. Additionally, they might now make a concerted effort to reestablish their image.

15-15 H_o : There is no difference between the two banks with respect to customer longevity.

H_a : There is a significant difference between the two banks with respect to customer longevity

α = .05

Given the data available, an appropriate test is the Kolmogorov-Smirnov two sample, two tailed test.

We use Table 15-7 in the text to find the critical D value. For alpha = .05:

$$\text{critical D} = 1.36 \sqrt{\frac{n_1 + n_2}{(n_1)(n_2)}}$$

$$\text{critical D} = 1.36 \sqrt{\frac{100 + 100}{(100)(100)}}$$

$$\text{critical D} = .192$$

Decision Rule: If: Calculated $|D| > .192$ we reject the null hypothesis, otherwise do not reject.

To find the calculated D we set up the following table:

	0-1	1-4	Years 4-10	Over 10	Total
Bank A Frequency	140	76	100	70	386
$S_1(x)$	140/386	216/386	316/386	386/386	
Bank B Frequency	221	80	55	40	396
$S_2(x)$	221/396	301/396	356/396	396/396	
$S_1(x) - S_2(x)$	-.195	-.200	-.080	0	

↑ D

Since: $|D| = |-.200| > .192$ our decision is to reject the null hypothesis and conclude there is a difference in the two banks.

15-17 The Mann-Whitney U test is used to test the following hypothesis:

$$H_o : \mu_1 = \mu_2$$

$$H_a : \mu_1 \neq \mu_2$$

$$\alpha = .10$$

Where μ_1 and μ_2 equal the true average premiums received by the two companies.

Decision Rule: Take a sample from each of the populations and determine Z:

$$\text{If} : -Z_{crit} \leq Z \leq Z_{crit}, \text{ accept } H_o$$

$$\begin{matrix} -Z_{crit} > Z \\ Z_{crit} < Z \end{matrix} \quad \text{reject } H_o$$

$Z_{critical}$ is shown in the following figure:

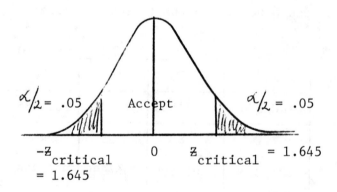

To deterimine the Z value we must complete the following steps:

1. Rank the observations after combining the two groups. Then find the sum of the two rankings as follows:

Company 1		Company 2	
Premium	Rank	Premium	Rank
246	5	300	8
211	2	305	9
235	3	308	10
270	6	325	14
411	18	340	16
310	11	295	7
450	19	320	13
502	20	330	15
311	12	240	4
200	1	360	17
	$\Sigma = 97$		$\Sigma = 113$

2. Calculate a U value for both samples:

$$U_1 = (n_1)(n_2) + \frac{n_1(n_1 + 1)}{2} - \Sigma R_1$$

$$= (10)(10) + \frac{10(10 + 1)}{2} - 97$$

$$= 100 + \frac{110}{2} - 97$$

$$= 58$$

$$U_2 = (n_1)(n_2) + \frac{n_2(n_2 + 1)}{2} - R_2$$

$$= (10)(10) + \frac{10(10 + 1)}{2} - 113$$

$$= 100 + \frac{110}{2} - 113$$

$$= 42$$

3. The U values will be approximately normally distributed with

$$\text{Mean} = \frac{(n_1)(n_2)}{2} = \frac{10 \times 10}{2} = 50$$

and

$$\text{Std Dev} = \sqrt{\frac{(n_1)(n_2)(n_1 + n_2 + 1)}{12}} = \sqrt{\frac{(10)(10)(10+10+1)}{12}}$$

$$= 13.23$$

4. The Z value can now be found as follows by selecting either U_1 or U_2

$$Z = \frac{U_1 - \text{Mean}}{\text{S.D.}}$$

$$Z = \frac{58 - 50}{13.23}$$

$$Z = .605$$

Since Z falls in the acceptance region, we will accept the null hypothesis and conclude there is no difference between the two regional companies, in terms of average premium size.

15-19 To analyze the data using the Kolmogorov-Smirnov Test the data should be presented into the following table.

	1100-1500	1501-1900	1901-2300	2301-2700	2701-3100
Frequency (Plan 1)	2	5	4	0	0
$S(x_1)$	2/11	7/11	11/11	11/11	11/11
Frequency (Plan 2)	0	1	5	3	2
$S(x_2)$	0/11	1/11	6/11	9/11	11/11
$\lvert S(x_1) - S(x_2) \rvert$	2/11 = .182	6/11 = .545	5/11 = .455	2/11 = .182	0 = 0.0

$$D = \max \lvert S(x_1) - S(x_2) \rvert$$

The hypothesis to be tested is:

H_o : There is no difference between advertising plans

H_a : There is a statistical difference between the advertising plans

Decision Rule: If: $D \leq D_{critical}$, accept H_o

$D > D_{critical}$, reject H_o

$\alpha = .05$

From Appendix 1-H: $D_{critical} = 1.36 \sqrt{\dfrac{11 + 11}{121}} = .5799$

Since the maximum difference in cumulative relative frequencies is less than $D_{critical}$, we cannot reject H_o.

You should note that in this case, the Kolmogorov-Smirnov two sample test leads to acceptance of the null hypothesis while in problem 16, using the Mann-Whitney U test we rejected the hypothesis with the same sample data. Normally, the Kolmogorov-Smirnov two sample test will lead to rejection whenever the Mann-Whitney test does. This points out the fact that when the data are grouped, the Kolmogorov-Smirnov is sensitive to the way the data are grouped.

15-21 The National Life Insurance Company is interested in determining if there is a difference between average annual premiums collected by the two companies they are considering for a merger. The hypothesis is:

$$H_o : \mu_1 = \mu_2$$

$$H_a : \mu_1 \neq \mu_2$$

$$\alpha = .10$$

Since the data are grouped into classes, an appropriate nonparametric test is the Kolmogorov-Smirnov two sample test. We begin by developing a cummulative relative frequency distribution table as follows:

Company 1	100-200	201-300	301-400	401-over
Frequency	7	21	40	2
Cumm. Rel. Freq.	7/70	28/70	68/70	70/70
$S(x_1)$	.10	.40	.971	1.00
Company 2				
Frequency	3	33	25	9
Cumm. Frequency	3/70	36/60	61/70	70/70
$S(x_2)$	.042	.514	.871	1.00
$S(x_1) - S(x_2)$	.058	-.114	.100	0

↑
D

Now we look to Appendix 1-H for the critical D level which is:

$$D_{critical} = 1.22 \sqrt{\frac{n_1 + n_2}{(n_1)(n_2)}}$$

$$D_{critical} = .2062$$

Since the calculated $D = |-.114| < .2062$, we must accept H_o based upon these data.

15-23 This problem can be solved using contingency table analysis. The expected, and observed, frequencies are shown in the following table.

	Taxes				
Work Hours	0-3000	3001-5000	5001-10000	over 10000	Σf_o
0 - 2	$f_o = 27$ $f_e = 21.09$	$f_o = 30$ $f_e = 25.74$	$f_o = 5$ $f_e = 10.73$	$f_o = 2$ $f^e = 6.44$	64
2 - 4	$f_o = 22$ $f_e = 20.77$	$f_o = 30$ $f^e = 25.35$	$f_o = 5$ $f^e = 10.56$	$f_o = 6$ $f^e = 6.34$	63
Over 4	$f_o = 10$ $f_e = 17.14$	$f_o = 12$ $f^e = 20.92$	$f_o = 20$ $f^e = 8.72$	$f_o = 10$ $f^e = 5.23$	52
Σf_o	59	72	30	18	179

The hypothesis is:

H_o : The work hours and taxes owed <u>are</u> independent.

H_a : The work hours and taxes owed <u>are not</u> independent.

Decision Rule: If: $\chi^2 \leq \chi^2_{critical}$, accept H_o

$\chi^2 > \chi^2_{critical}$, reject H_o

$\alpha = .10$

χ^2 critical is shown in the following figure:

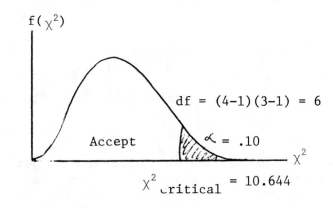

$$\chi^2 = \sum_{i=1}^{row} \sum_{j=1}^{col} \frac{(f_o - f_e)^2}{f_e}$$

χ^2 is found below:

Row	Column	f_o	f_e	$(f_o-f_e)^2$	$(f_o-f_e)^2/f_e$
1	1	27	21.09	34.93	1.66
1	2	30	25.74	18.15	.71
1	3	5	10.73	32.84	3.06
1	4	2	6.44	19.72	3.06
2	1	22	20.77	1.51	.01
2	2	30	25.34	21.72	.86
2	3	5	10.56	30.91	2.93
2	4	6	6.34	.12	.02
3	1	10	17.14	50.98	2.97
3	2	12	20.92	79.57	3.80
3	3	20	8.72	127.24	14.59
3	4	10	5.23	22.75	4.35

$$\chi^2 = 38.02$$

Since 38.02 > 10.644, we will reject the null hypothesis and conclude that a relationship exists between the time it takes to complete a return and the taxes owed by the client.

CHAPTER 16

SIMPLE LINEAR REGRESSION AND CORRELATION ANALYSIS

The previous chapters in this text have dealt with various means of analysis involving a single variable. However, as a decision maker, you will encounter many instances in which it will be important to describe the relationship between two or more variables. For example, a financial consultant may wish to determine the relationship between a company's earnings and various other variables such as sales, administrative overhead, market share, and so forth.

Two very useful statistical tools for helping the decision maker describe the relationship between two or more variables are correlation and regression analysis.

16-1 Statistical Relationships Between Two Variables:

Two random variables (Y and X) can be related in one of the following ways:

1. Linear relationship (positive or negative)

2. Curvilinear relationship (positive or negative)

3. No relationship

If two variables are linearly related, a change in X will be accompanied by a systematic change in the Y variable. A scatter plot of two linearly related variables will show a straight line relationship between them.

If two variables are related in a curvilinear manner, the change in the Y variable is not constantly proportional to the change in the X variable. However, the relationship is recognizable.

If two variables are not related, the scatter plot will show no recognizable pattern. That is when X increases, sometimes Y increases, and sometimes Y decreases.

When we analyze the relationship between two variables, we generally call one variable the dependent variable and the other the independent variable. The dependent variable is the variable whose variation we try to explain. The independent variable is used to explain the variation in the dependent variable.

16-2 Correlation Analysis:

The correlation coeffficient is the quantitative measure of strength in the linear relationship between any two variables. The correlation coefficient is computed using:

$$r = \frac{n\sum XY - \sum X \sum Y}{\sqrt{(n\sum X^2 - (\sum X)^2)(n\sum Y^2 - (\sum Y)^2)}}$$

where

r = correlation coefficient

n = sample size

X = value of the independent variable

Y = value of the dependent variable

The correlation between two variables can range from +1.0 (perfect positive linear relationship) to -1.0 (perfect negative linear relationship). A correlation coefficient of 0.0 indicates no linear relationship.

We can test the statistical significance of the correlation coefficient by the following t statistic:

$$H_o \;\; \rho = 0.0$$

$$H_A \;\; \rho \neq 0.0$$

$$t = \frac{r}{\sqrt{(1 - r^2)/(n - 2)}}$$

where

r = correlation coefficient

n = sample size

n - 2 = degrees of freedom associated with t statistic from the t table in the text.

A word of caution! Just because two variables are determined to be statistically correlated, does not mean that a change in one variable causes the other to change. There can be no cause and effect interpretation based only on the knowledge that two variables are correlated.

16-3 __Simple Coefficient of Determination:__

The __coefficient of determination__, r^2, measures the percentage of variation in the Y variable which can be explained by the variation in the X variable. It is found by squaring the correlation coefficient.

$$\text{Simple coefficient of determination} = r^2$$

Thus, if $r = .50$, the percentage of the variation explained is $r^2 = (.50)^2 = .25$. Note, the close the correlation gets to ± 1.0, the closer r^2 gets to 1.0.

The statistical significance of the coefficient of determination is tested as follows:

$$H_o \; \rho^2 = 0.0$$

$$H_A \; \rho^2 \neq 0.0$$

$$F = \frac{r^2}{(1 - r^2)/(n - 2)}$$

Note that this test is an F test with 1 and $n - 2$ degrees of freedom. Large values of F will lead to rejection of the null hypothesis implying a significant percentage of explained variation.

16-4 __Simple Linear Regression Analysis:__

The basic form of the simple regression model is:

$$Y_i = B_o + B_1 X_1 + e_i$$

where

Y_i = dependent variable

X_i = independent variable

B_o = Y intercept

B_1 = slope of the regression line

e_i = residual or random error

This regression model has 4 basic assumptions:

1. Values of the Y variable are statistically independent

2. For any X, there exist many possible values of Y whose

distribution is normal

3. The distributions of the possible Y values have the same variance for all values of X

4. The average value of Y at each value of X can be connected by a straight line called the population regression line (model)

The slope, B_1, represents the average change in Y for each unit change in X. The slope can be positive or negative depending on the direction of the relationship between Y and X.

The intercept, B_o, is the mean value of Y when X = 0. The intercept has meaning only when the data contain cases with X = 0.

16-5 Estimating the Simple Regression Model--The Least Squares Approach:

Many possible regression lines could be fit to a set of data. In order to settle on one "best" regression line, we must adopt a criterion. The criterion generally used is known as the <u>least squares criterion</u>. The least squares criterion selects as the "best" regression line the one which minimizes the following:

$$SSE = \sum_{i=1}^{n} (Y_i - \hat{Y}_i)^2$$

where

SSE = sum of squares error

Y = actual value of the dependent variable

$\hat{Y}$ = estimated value of the dependent variable

Note, the $\hat{Y}$ values are derived from the sample regression line for the sample data. This sample regression line has the following form:

$$\hat{Y}_i = b_o + b_1 X_i$$

where

b_o = estimated Y intercept

b_1 = estimated slope coefficient

The sample regression coefficients, b_o and b_1 which minimize SSE are calculated by:

$$b_1 = \frac{\sum XY - \frac{(\sum X)(\sum Y)}{n}}{\sum X^2 - \frac{(\sum X)^2}{n}}$$

and

$$b_0 = \overline{Y} - b_1 \overline{X}$$

When the slope and intercept are computed using the above equations, b_0 and b_1 are called the least squares estimates of B_0 and B_1 respectively. The least squares regression criteria has the following properties associated with it:

1. The sum of the residuals is zero.

$$\sum (Y - \widetilde{Y}) = 0$$

2. The sum of the squared residuals is minimized.

$$\sum (Y - \widetilde{Y})^2 \text{ is minimum}$$

3. The simple regression line passes through the point $\overline{X}, \overline{Y}$.

4. On the average, b_0 and b_1 will equal B_0 and B_1 respectively. The least squares estimates are <u>unbiased</u>.

16-6 Significance Tests in Regression Analysis:

The total variation to be explained in the dependent variable is:

$$SST = \sum (Y - \overline{Y})^2$$

This SST can be partitioned into that which the regression model explains, SSR, and that which is left unexplained, SSE. We know that SSR = SST - SSE. Therefore, for the sample data, the percentage of total variation explained by the model (called the coefficient of determination) is:

$$r^2 = \frac{SSR}{SST}$$

If the regression model is significant, the ratio of explained variation to unexplained variation will be "large." Likewise r^2 will be "large." Analysis of variance can be used to test the significance of the regression as follows:

H_o : model does not explain any variation in the dependent variable

H_A : model does explain variation in the dependent variable

ANOVA

Source	SS	DF	MS	F
Regression	SSR	1	SSR/1	$\frac{SSR/1}{SSE/n-2}$
Unexplained (Error)	SSE	n-2	SSE/n-2	
Total	SST	n-1		

If the calculated F exceeds the appropriate F value with 1 and n-2 degrees of freedom, we reject H_o.

Significance of the Slope:

A parallel test of significance to ANOVA test is the test that the true regression slope is zero.

The variation in possible sample regression slopes is estimated by:

$$S_{b_1} = \frac{\sqrt{\frac{SSE}{n-2}}}{\sqrt{\Sigma(X - \overline{X})^2}}$$

Then the significance test for the regression slope is:

$$H_o : B_1 = 0$$

$$H_A : B_1 \neq 0$$

$$t = \frac{b_1 - B_1}{S_{b_1}}$$

A large t will lead to rejection of the null hypothesis since $b_1 - B_1$ will be large relative to S_{b_1}.

16-7 Regression Analysis for Prediction:

One of the main uses of regression analysis is prediction. We establish a model of the form:

$$\hat{Y} = b_o + b_1 X$$

Then we select values of X and substitute into the equation to obtain $\hat{y}$, a prediction of the true value of Y. Note, the least squares model was developed to "best" fit the original data. However, there is no reason to believe that the prediction will offer as good a fit to the true value. This is especially true if the X value is outside the range of relevant data. Care should always be used when extrapolating beyond the relevant range.

The value $\hat{Y}$ is a <u>point estimate</u> and is subject to sampling error. As with estimates of the population mean, we can develop confidence interval estimates for Y.

The intervals are of two forms:

1. Predict average Y|X:

$$\hat{Y} \pm t \sqrt{\frac{SSE}{n-2}} \sqrt{\frac{1}{n} + \frac{(X_p - \bar{X})^2}{\sum(X - \bar{X})^2}}$$

where

$$X_p = \text{value of X used to find } \hat{Y}$$

Note, as X_p departs in either direction from $\bar{X}$, the interval becomes wider (less precise).

2. Predict particular Y|X:

$$\hat{Y} \pm t \sqrt{\frac{SSE}{n-2}} \sqrt{1 + \frac{1}{n} + \frac{(X_p - \bar{X})^2}{\sum(X - \bar{X})^2}}$$

By examining these two confidence interval formulas we see that the interval generated for predicting a particular Y|X will always be less precise than for predicting an average Y|X.

16-8 Regression Analysis for Description:

Although regression analysis has many applications in the area of prediction, other applications involve description. Here the decision maker is more concerned with identifying and describing relationships between variables. Such things as the significance, size, and sign of the regression slope coefficient are examined to obtain information or test a theory about the estimated change in the dependent variable for a unit change in the independent variable.

16-9 Regression Analysis for Control:

Control-related applications are probably less numerous than prediction and description applications, however, regression can be a valuable tool for control purposes. If the recommended increase in sales revenue for each dollar increase in advertising is supposed to be a certain level, say $10, for a particular product, sample data can be collected and a regression model of the form

$$\text{Sales} = b_o + b_1 (\text{Advertising})$$

can be developed. Then if b_1 departs significantly from $10 we might conclude something is either wrong with our advertising or it is doing much better than expected, depending on the direction of the difference.

16-10 <u>Conclusions</u>:

Regression analysis is heavily used by business decision makers to analyze the relationship between two or more variables for applications involving prediction, description, and control.

Correlation measures the strength of the linear relationship between two variables but cannot be used to imply cause and effect. However, if a dependent and independent variable are highly correlated (close to ± 1.0), the resulting regression model will tend to provide a close fit to the observed data. In these instances the coefficient of determination will be close to 1.0.

Regression analysis, while a powerful tool, can be misused. For example, if predictions are made based on observations of the independent variable, X, outside the range of relevant data, the predictions may be very misleading.

```
*******************************
*                             *
*                             *
*          SOLUTIONS          *
*                             *
*******************************
```

16-1   (a) Coefficient of Correlation. This value is a measure of the linear relationship between two variables. The linear relationship can be positive with +1.0 indicating a perfect linear relationship or negative with -1.0 indicating a perfect negative linear relationship. The closer r gets to 0, the weaker the linear relationship becomes.

(b) Scatter Diagram. A scatter diagram is a two-dimensional plot showing one variable on the horizontal axis and the second on the vertical axis. The plot of points provides a visual indication of the relationship between the two variables.

(c) Least Squares Criterion. In simple regression analysis, a line is fit to a set of points such that it is the "best" fit. But best by what criterion? It is generally accepted that the "best" line is the one which minimizes the sum of squared errors. That is the least squares line minimizes $\Sigma(Y - \hat{Y})^2$.

(d) Estimates of Regression Coefficients. If in fact the entire population of cases were available, the regression model which would be developed would be the "parameter" model. The slope and intercept coefficients would be the true slope and intercept values. However, in most cases, the decision maker has only a sample of cases from the population. From that sample he or she can calculate the least squares regression model for the sample data. The least squares equations are:

$$b_1 = \frac{\Sigma XY - \frac{(\Sigma X)(\Sigma Y)}{n}}{\Sigma X^2 - \frac{(\Sigma X)^2}{n}}$$

and

$$b_0 = \bar{Y} - b_1 \bar{X}$$

These coefficients will, on the average, equal the true coefficients for the population.

(e) Confidence Interval for a Regression Estimate. As with any sampling environment, the regression coefficients are subject to sampling error. In addition, the variability in the dependent variable is composed of two sources; that which can be explained by the independent variable and the unexplained variation. Because of sampling error and because of the unexplained variation, a point estimate of dependent variable based upon the regression model is also subject to error. A confidence interval estimate provides a range (or interval) within which the decision maker can be confident that the true value of the dependent variable lies.

16-3   (a)

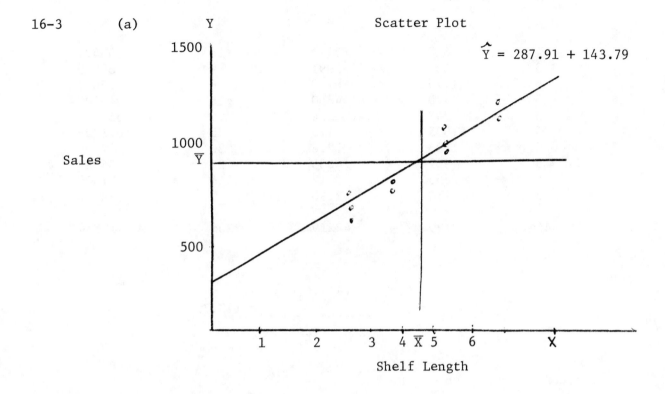

Scatter Plot

$\hat{Y} = 287.91 + 143.79$

Shelf Length

Based upon this scatter plot, it would appear that there is a strong linear relationship between the sales and shelf space allocated to the cosmetic brand. However, such a plot does not infer a cause-and-effect relationship between the two variables.

(b) To determine the least squares regression line, we must find $b_o + b_1$ as follows:

$$b_1 = \frac{\sum XY - \frac{(\sum X)(\sum Y)}{n}}{\sum X^2 - \frac{(\sum X)^2}{n}}$$

$$b_o = \overline{Y} - b_1 \overline{X}$$

X	Y	XY	X^2	Y^2
4	868	3472	16	753424
3	697	2091	9	485809
6	1125	6750	36	1265625
5	970	4850	25	940900
3	742	2226	9	550564
5	1035	5175	25	1071225
6	1203	7218	36	1447209
5	967	4835	25	935089
4	853	3412	16	727609
3	730	2190	9	532900
44	9190	42219	206	8710354

Thus:

$$b_1 = \frac{42219 - \frac{(44)(9190)}{10}}{206 - \frac{(44)^2}{10}} = 143.79$$

$$b_o = 287.91$$

The least squares line is:

$$\hat{Y} = 287.91 + 143.79(X)$$

See line plotted on scatter plot. Note, the line passes through the point $\bar{Y}$ and $\bar{X}$.

16-5 Let Xp = 5

Then

$$\hat{Y} = 287.91 + 143.79(X_p = 5)$$

$$\hat{Y} = 1006.86$$

Thus, the point estimate for sales given 5 feet of shelf space is $1006.86.

Like any other point estimate we cannot place any confidence in this estimate. Consequently, we will develop a 95% confidence interval for the average sales level given five feet of shelf space as follows:

$$\hat{Y} \pm t_{\alpha/2, 8} \sqrt{\frac{SSE}{n-2}} \sqrt{\frac{1}{n} + \frac{(X_p - \bar{X})^2}{\sum (X_i - \bar{X})^2}}$$

where SEE is computed as follows:

X	Y(Actual)	$\tilde{Y}$(Estimated)	(Y-$\hat{Y}$)Error	$(Y-\hat{Y})^2$
4	868	863.07	4.93	24.3049
3	697	719.28	-22.28	496.3984
6	1125	1150.65	-25.65	657.9225
5	970	1006.86	-36.86	1358.6596
3	742	719.28	22.72	516.1984
5	1035	1006.86	28.14	791.8596
6	1203	1150.65	52.35	2740.5225
5	967	1006.86	-39.86	1588.8196
4	853	863.07	-10.07	101.4049
3	730	719.28	10.72	114.9184

$$\Sigma = 8391.0088 = SSE$$

$$SEE = \sqrt{\frac{\Sigma(Y-\hat{Y})^2}{n-2}} = \sqrt{\frac{8391.0088}{8}} = 32.33$$

$$1006.86 \pm (2.306)(32.33)\sqrt{\frac{1}{10} + \frac{(5-4.4)^2}{12.40}}$$

$$1006.86 \pm 26.78$$

$$\$980.08 \text{ ———— } \$1033.64$$

The 95% confidence interval for the particular sales level given the 5 feet of shelf space is:

$$Y \pm t_{\alpha/2, 8} \sqrt{\frac{SSE}{n-2}} \sqrt{1 + \frac{1}{n} + \frac{(X_p - \overline{X})^2}{\Sigma(X-\overline{X})^2}}$$

$$1006.86 \pm 2.306(32.33)\sqrt{1 + \frac{1}{n} + \frac{(5-4.4)^2}{12.40}}$$

$$1006.86 \pm 79.21$$

$$927.65 \text{ ———— } 1086.07$$

16-7 The intercept (b_o) of a regression model has meaning only when the values of the independent variable include zero. If the X values do not include zero, no particular meaning can be attached to the intercept coefficient. Since, in the sample, the manager has never tried to sell cosmetics with zero shelf space, it is improper to try to attach meaning to the intercept.

A regression estimate based upon shelf length of 20 feet would be tenuous at best since the model was not developed with any space as great as 20 feet. You must be very careful when extending beyond the range of data because the relationship between X and Y may change.

16-9 (a) We have used a computer program to perform the necessary calculations. We suggest you consult the methodology of solutions 16-4 through 16-8 if you have trouble with manual procedures in solving this problem and problems 16-10 through 16-13.

$$Y = \text{sales (in millions)}$$

$$X_1 = \text{income current dollars}$$

$$X_2 = \text{income 1967 dollars}$$

The regression models will be of the form:

$$\hat{Y} = b_0 + b_1(X_i)$$

For the regression using X_1 (income in current dollars) we get:

$$\hat{Y} = 166.46 + .0537(X_1)$$

and for X_2 (income in 1967 dollars) we get:

$$\hat{Y} = 239.66 + .0639(X_2)$$

(b) and (c)

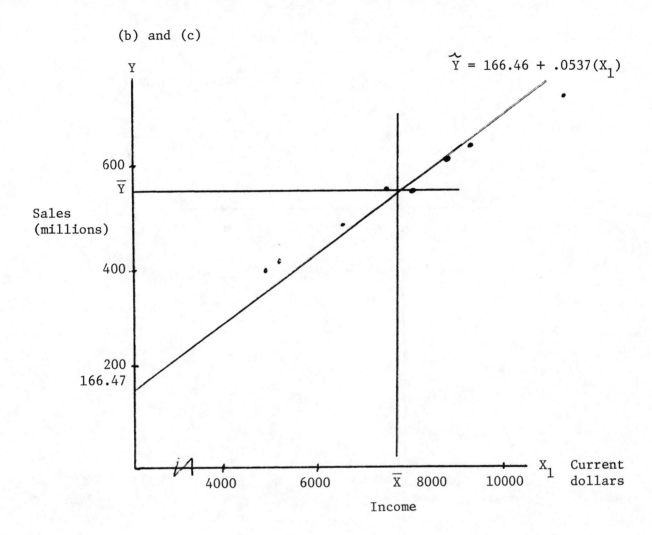

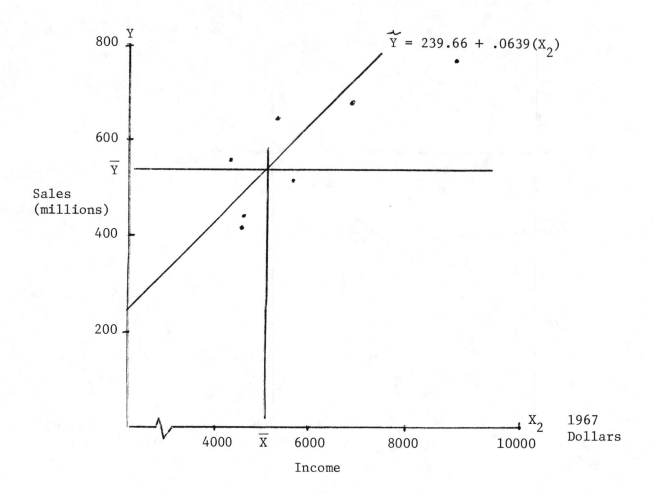

Based upon the scatter plots and the fitted least squares lines, the model which appears to provide the better explanation of sales variability is the one using farm family income in current dollars. The regression line appears to better fit the data points. That is, the points appear to follow a linear pattern more closely.

16-11   Using the model:

$$\hat{Y} = 166.462 + .0537(X_1)$$

$$\hat{Y} = 166.462 + .0537(8130)$$

$$\hat{Y} = 603.04 \text{ is the point estimate given the regression model developed in problem 16-9}$$

Like any point estimate, the decision maker cannot place any confidence in the estimated sales value.

To develop a 98% confidence interval for the average value of sales given current dollars income of $8,130:

$$\hat{Y} \pm t_{\alpha/2,(n-2)} \text{ SEE } \sqrt{\frac{1}{n} + \frac{(X_p - \bar{X})^2}{\Sigma X_i^2 - \frac{(\Sigma X_i)^2}{n}}}$$

Again, using computer generated output:

$$\hat{Y} \pm (3.143)(16.48) \sqrt{\frac{1}{8} + \frac{(8130 - 7624.12)^2}{36,980,760}}$$

$$\hat{Y} \pm 18.81$$

The 98% confidence interval is:

603.04 ± 18.81

584.23 ——————— 621.85

16-13 The format for a confidence interval for the regression slope is:

$$b_1 \pm t_{\alpha/2(n-2)} S_{b_1}$$

where

$$S_{b_1} = \frac{\sqrt{\frac{SSE}{n-k-1}}}{\sqrt{\Sigma X_i^2 - \frac{(\Sigma X)^2}{n}}}$$

Thus: for Model 1 using computer generated values

95% confidence interval:

.0537 ± (2.447)(.00272)

.0537 ± .00665

.04705 ——————— .06035

For Model 2

95% confidence interval:

$$.0639 \pm (2.447)(.01779)$$

$$.0639 \pm .0434$$

$$.0204 \text{ ———— } .1074$$

16-15 There are at least two equivalent approaches for discussing the statement that TV hours is not a significant explanation of the variation in student grade point average.

First approach:

Test H_o $\rho = 0.0$

H_A $\rho \neq 0.0$

$\alpha = .05$

Test statistic:

$$t = \frac{r\sqrt{n-2}}{\sqrt{1-r^2}} = \frac{(-.4926)\sqrt{48}}{\sqrt{1-(.4926)^2}}$$

$$t = -3.926$$

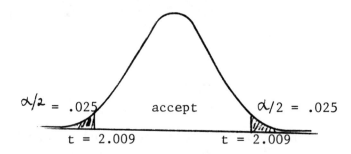

$\alpha/2 = .025$ accept $\alpha/2 = .025$
$t = 2.009$ $t = 2.009$

Decision Rule:

If: calculated $t > 2.009$ or < -2.009 reject H_o
otherwise, do not reject

Since $t = -3.926 < -2.009$ we reject H_o and conclude that TV hours is linearly related to grade point average at α .05 level.

Second approach:

H_o $B_1 = 0$

H_A $B_1 \neq 0$

$\alpha = .05$

Test statistic:

$$t = \frac{\hat{B}_1 - 0}{S_{\hat{B}_1}} = \frac{-.0015 - 0}{.000382} = -3.926$$

Use same decision rule as approach 1. Since t = -3.926 < -2.009 reject.

16-17 Points to consider:

1. Cause and effect cannot be assumed <u>just because</u> two variables are <u>linearly</u> correlated.

2. Use care in extending the use of X values outside the range of data used in developing the model.

3. There is some maximum GPA (probably 4.0). The statement would indicate that GPA can increase forever, simply by increasing hours worked.

4. While the correlation is significantly greater than 0.0 it is not high. The low r^2 would indicate that hours worked explains little of the variation in grade point average.

5. Refute Henry's statement.

16-19 The slope coefficient measures the average change in Y for a one-unit change in X. For this model, the regression slope estimate is -.0015. This means that on average, a one-unit increase in TV hours is expected to produce an average decline in grade point average of .0015 points.

The $\hat{B}_1$ is a point estimate. We can place no confidence in it. To develop a 90% confidence interval for the slope coefficient, we use the following format:

$$\hat{B}_1 \pm t_{\alpha/2, n-2} (S_{\hat{b}_1})$$

$$-.0015 \pm 1.676 \ (.000382)$$

$$-.0015 \pm .00064$$

$$-.00214 \text{ ———————— } -.00084$$

Of all possible confidence intervals for the slope based upon a sample of size 50, approximately 90% will include the true slope, B_1. We can be fairly sure that the above interval includes B_1.

16-21 (b) To develop the least squares regression line, we set up the following:

X	Y	XY	X^2	Y^2
83	90	7470	6889	8100
57	60	3420	3249	3600
80	72	5760	6400	5184
69	77	5313	4761	5929
95	96	9120	9025	9216
84	90	7560	7056	8100
84	79	6636	7056	6241
74	65	4810	5476	4225
84	80	6720	7056	6400
72	75	5400	5184	5625
782	784	62209	62152	62620

Then:

$$b_1 = \frac{\sum XY - \frac{\sum X \sum Y}{n}}{\sum X^2 - \frac{(\sum X)^2}{n}}$$

$$b_1 = \frac{62209 - \frac{(782)(784)}{10}}{(62152) - \frac{(782)^2}{10}} = \frac{900.2}{999.6}$$

$$b_1 = .9002$$

and

$$b_o = \bar{Y} - b_1 \bar{X}$$

$$b_o = 78.4 - (.9002)(78.2)$$

$$b_o = 8.004$$

Thus, the least squares regression model from these data is:

$$\hat{Y} = 8.004 + .9002 \text{ (students average)}$$

16-23 The format for a 95% confidence interval for estimating a particular student's grade in the class given a 95 average in other classes is:

$$\hat{Y} \pm t_{\alpha/2} \text{ SEE } \sqrt{1 + \frac{1}{n} + \frac{(X_p - \bar{X})^2}{\Sigma(X_i - \bar{X})^2}}$$

where

$$Y = 8.004 + (.9002)(95)$$

$$Y = 93.52$$

We also compute SEE as follows:

X	Y(actual)	$\hat{Y}$(estimated)	$Y-\hat{Y}$	$(Y-\hat{Y})^2$
83	90	82.72	7.28	52.99
57	60	59.31	.69	.48
80	72	80.02	-8.02	64.32
69	77	70.11	6.89	47.47
95	96	93.52	2.48	6.15
84	90	83.62	6.38	40.70
84	79	83.62	-4.62	21.34
74	65	74.61	-9.61	92.35
84	80	83.62	-3.62	13.10
72	75	72.83	2.17	4.71
				343.61

Thus:

$$\text{SEE} = \sqrt{\frac{\Sigma(Y-\hat{Y})^2}{n-2}} = \sqrt{\frac{343.61}{8}} = 6.55$$

and

$$\Sigma(X-\bar{X})^2 = \Sigma X^2 - \frac{(\Sigma X)^2}{n} = 62152 - \frac{(782)^2}{10}$$

$$\Sigma(X-\bar{X})^2 = 999.6$$

Thus the 95% confidence interval is:

$$93.52 \pm (2.306)(6.55) \sqrt{1 + \frac{1}{10} + \frac{(95 - 78.2)^2}{999.6}}$$

$$93.52 \pm 17.75$$

$$75.77 \text{ ——————— } 111.25$$

These results indicate that we have a high level of confidence (95%) that the true grade in the class for a particular student will be between 75.77 and 111.25 given an overall average of 95.

16-25 It is clear that the precision of interval developed to estimate a particular student's score is lower (wider interval) than the precision of the interval developed to estimate the average score for all students who have the same particular average score in other classes. This should make intuitive sense because the variation in individual scores will be greater than for the average of individuals. That is, while we may observe an unusually high or low score by an individual, the average of several students will be pulled toward the center.

16-27 The correlation between fuel consumption and number of rail cars is:

$$r = \frac{n \sum X Y - \sum X \sum Y}{\sqrt{(n \sum X^2 - (\sum X)^2)(n \sum Y^2 - (\sum Y)^2)}}$$

We have solved this problem using the SPSS computer program. The computed correlation is:

$$r = .95065$$

The high correlation coefficient confirms our visual assumption in problem 26 that there appears to be a strong linear relationship between fuel consumption and the number of railroad cars on the train. However, we must not assume a cause-and-effect situation is present. A high correlation can <u>never</u> be assumed to imply cause-and-effect by itself.

16-29 The measure of variation in the dependent variable explained by the independent variable is called the coefficient of determination or R^2.

$$R^2 = \frac{SSR}{SST}$$

where:

SST = sum of squares total

SST = SSR + SSE

and

SSR = sum of squares regression

SSR = SST - SSE

and

SSE = sum of squares error

$$SSE = (Y - \hat{Y})^2$$

We have used a computer program to solve for R^2.

$$R^2 = \frac{4726.20}{5229.59} = .90374$$

This indicates that slightly over 90% of the variation in fuel consumption can be explained by the number of rail cars on the train.

16-31 This problem asks us to develop a 95% confidence interval estimate of the average change in fuel costs for each additional rail car added to the train.

Recall from problem 28 and 30 that the regression model is:

$$\hat{Y} = 10.481 + 2.154(X)$$

Where:

$\hat{Y}$ = fuel units used

X = rail cars

2.154 = slope coefficient

10.481 = intercept (constant)

The point estimate for the average change in fuel units for each additional rail car is the least squares slope coefficient, 2.154. The format for a 95% confidence interval for the slope is:

$$b_1 \pm t_{\alpha/2} \, S_{b_1}$$

Using the information computed from problem 28 we get:

$2.154 \pm (2.306)(.248)$

$2.154 \pm .571$

1.583 units —————— 2.725 units

Now to transform this interval estimate into an estimate for the change in dollar cost per rail car added, we multiply the lower and upper limits by $10.00. Thus, the appropriate interval estimate is:

$15.83 —————— $27.25

CHAPTER 17

INTRODUCTION TO MULTIPLE REGRESSION ANALYSIS

In Chapter 16 we discussed regression analysis as a common method of analyzing the relationship between two variables. While this simplified discussion forms a necessary starting point, simple regression analysis will satisfy only some practical managerial applications. In many other cases, the operational manager is considering the relationship between many variables. Fortunately, we can extend the concepts discussed in the previous chapter quite easily in the form of multiple regression analysis.

17-1 A Non-Quantitative Analogy for Multiple Regression Analysis:

Often managers are literally overwhelmed with the number of possible variables that might be considered in attempting to explain the variation in a dependent variable of interest. However, if the manager is going to use several variables in this decision-making situation, he will be interested not only in choosing the independent variables that are individually effective, but in also choosing independent variables that are effective when working together in explaining the dependent variable. Just as many jobs are most effectively performed by a team, many decision-making situations are best analyzed by a team of variables. Likewise, just as effective team play becomes a important criterion in hiring a worker, decision variables are also judged on how well they interact with the other available variables.

17-2 The Multiple Regression Model:

When dealing with a multiple regression model we divide the variables into two groups:

1. The dependent variable: the variable we are interested in being able to predict or variable whose variation we seek to explain.

2. The independent variables: all those variables we can use to predict, or explain the variation in the dependent variable.

Which variable is the dependent variable depends on our application. Likewise the independent variables used depend on their availability and relationship with the dependent variable and each other.

The value of our regression model depends on how good the group of independent variables is in modeling the changes in the dependent variable.

The general criterion is to find a set of independent variables that are highly related to the dependent variable and not related to each other. This relationship is generally found by looking at the correlation between all sets of variables. We would like to have a set of independent variables highly correlated with the dependent variable and not correlated with each other. If the independent variables are too highly correlated with each other, we may have problems with our multiple regression model. These problems come under the heading of multicolinearity.

Once we have determined an appropriate set of independent variables, we are able to construct a multiple regression model of the following form:

$$Y_i = B_o + B_1 X_{1i} + B_2 X_{2i} + \ldots + B_k X_{ki} + e_i$$

where:

Y_i = ith observation of the dependent variable

B_o = regression constant

B_i = regression slope coefficient for variable X_i

.
.
.

k = number of independent variables

e_i = random error

As you can see, this is an extension of the simple regression model of the last chapter. The following assumptions are made:

1. The random errors are normally distributed;

2. The mean of the random error terms is zero; and

3. The error terms have a constant variance for all combinations of values of the independent variables.

The major difference between a multiple regression model and the simple regression model is the number of independent variables the model conttains. Further, the mathematics of computing the regression coefficients make it impractical to work by hand. Many computer routines exist to do this.

17-3 <u>Applying the Regression Model</u>:

In using the regression model, the first task is to identify the independent variables that will help explain the variation in the dependent variable. Once these variables are identified, values for each variable are recorded for a number of observations (n). After the set of observations are recorded, the correlation coefficient between all combinations of the variables (both dependent and independent) is found using the following equation:

$$r = \frac{n \sum X_1 X_2 - \sum X_1 \sum X_2}{\sqrt{[n(\sum X_1^2) - (\sum X_1)^2][n(\sum X_2^2) - (\sum X_2)^2]}}$$

These correlation coefficients are arranged in a table of correlations called a correlation matrix (see Table 17-3 in the text).

High correlation coefficients between the dependent variable and an independent variable identify independent variables that will likely be useful in the regression model. High correlation coefficients between two independent variables signal possible future trouble for the regression model.

Analyzing the Regression Model

Once the appropriate set of independent variables have been identified and the model determined using a computer program, the decision maker is interested in answers to the following questions:

1. Is the overall model significant?

2. Which independent variables are significant?

3. Is the standard error of the estimate too large for the model to be useful?

4. Do the signs on the model's coefficients make sense?

We address each of these:

1. The first question is answered using the model's R^2 value where:

$$R^2 = \frac{\text{Sum of Squares Regression}}{\text{Sum of Squares Total}} = \frac{SSR}{SST}$$

This value indicates the proportion of variation in the dependent variable that can be explained using the regression model. This value can be tested using the F statistic:

$$F = \frac{SSR/K}{SSE/(n-K-1)}$$

Large values of the F statistic lead to rejection of the hypothesis that the regression model has <u>no</u> value in explaining the variation in the dependent variable.

Additional information is provided by R_A^2, the model's adjusted R^2. The adjusted R^2 takes into consideration that additional amounts of explained variation are gained at the expense of lost degrees of freedom in the model.

2. While the overall regression model may be significant, not all the independent variables will necessarily be significant in explaining variation in the dependent variable. Each independent variable can be tested using a t test. This t value, given on most computer printouts is computed by:

$$t = \frac{b_i - B_i}{S_{b_i}}$$

where:

b_i = sample regression coefficient

B_i = hypothesize slope (usually $B_i = 0$)

S_{b_i} = standard error of slope

3. In many practical decision-making situations, the answer to the third question must be based upon the decision maker's judgment. The Standard Error of the Estimate (SEE) is examined. If this value is acceptable to the decision maker, the model is useful. If this value is too large for the decision maker, the model will not be useful in a practical sense, even though it may be highly significant using an F test.

4. In many instances, the decision maker will have reason to believe that the regression coefficient sign should be positive (or negative) based upon his/her past experience or simply based upon theories and principles. For instance, the relationship between price of an item and the quantity demanded is an inverse relationship. If the regression coefficient had a positive sign, there would be reason to believe that either this set of data is unusual or there is a multicolinearity problem.

Coefficients that have unreasonable signs are often caused by multicolinearity. This occurs when the independent variables have high intercorrelation. Multicolinearity is best prevented by eliminating one of the two correlated independent variables from the model.

17-4 <u>Dummy Variables in a Regression Model</u>:

Many times variation in a dependent variable can be explained by categorical type variables. For instance, many studies have indicated that sex (male or female) can help explain variation in personal earning levels.

Categorical variables, like sex, can assume only certain distinct states. These variables can be incorporated into the regression model by using <u>dummy variables</u>. A dummy variable for a variable that can take on only two states is established as follows:

$$\text{Dummy variable} = X_6 = 1 \text{ if male}$$

$$\text{Dummy variable} = X_6 = 0 \text{ if female}$$

If the categorical variable has more than two levels, we have to use more than one dummy variable. For example, marital status might have the responses: single, married, divorced, other. Since there are four levels of response, we can create 3 dummy variables as follows:

$$X_1 = 1 \text{ if single}$$
$$X_1 = 0 \text{ otherwise}$$

$$X_2 = \text{if married}$$
$$X_2 = 0 \text{ otherwise}$$

$$X_3 = 1 \text{ if divorced}$$
$$X_3 = 0 \text{ otherwise}$$

Note, we need only 3 dummy variables in this case since the responses to these three automatically inform us whether the individual indicated "other" as his/her marital category.

Remember, always establish one fewer dummy variables than there are levels of response in the categorical variable of interest. Note also that categorical variables with large numbers of possible responses require the creation of many dummy variables.

17-5 <u>Stepwise Regression Models</u>:

Multiple regression models fall into two general types: those that include all possible independent variables in the analysis at one time; and those that bring variables into the model in a stepwise fashion. Stepwise regression models have the advantage of letting the decision maker analyze the regression model as it is being constructed. The stepwise models will indicate, on an incremental basis, which independent variables add most to explaining variation in the dependent variable. However, this may lead to some confusion about the overall effect of some of the independent variables. Both types of regression model will have the same levels of significance with the same independent variables included in the model.

17-6 Conclusions:

Multiple regression analysis is an extension of simple regression analysis. Rather than use one independent variable to help explain the dependent variable, in multiple regression we search for several independent variables which together can explain a significant proportion of the variation in the dependent variable.

Multiple regression is very heavily used as a statistical tool by business decision makers. Applications in the areas of prediction, description, and control occur frequently in all the functional business areas. The computer is used almost exclusively in performing the necessary computations in multiple regression applications. Chapter 17 has emphasized the importance of interpreting the computer output and testing the significance of the regression model and the individual variables included.

```
*****************************
*                           *
*                           *
*         SOLUTIONS         *
*                           *
*                           *
*****************************
```

17-1 (a) The simple linear model utilizes one independent or explanatory variable and one dependent variable. The multiple regression model utilizes two or more explanatory variables and a single dependent variable.

While there certainly are instances in which the simple linear model would be quite appropriate for business decision making, the use of multiple regression is more extensive. In many instances, more than one independent variable is required to adequately explain the variation in the dependent variable.

(b) The slope coefficients in a multiple regression model have the same basic interpretation as for the simple regression model. That is, a particular slope coefficient measures the average change in the dependent variable for a one-unit change in the independent variable holding all other variables in the model constant.

(c) The coefficient of multiple determination measures the percentage of variation in the dependent variable which has been explained by the independent variables in the model.

17-3   Let X_1 = Case Analysis Score

X_2 = Written Presentation Score

X_3 = Verbal Presentation Score

Y = Job Rating

The resulting regression model is:

$$\hat{Y} = 21.4805 + 2.364(X_1) + 1.531(X_2) + 3.807(X_3)$$

17-5 We can test the significance of the overall model by using the analysis of variance F test as follows:

H_o : all slope coefficients equal zero

H_A : not all slope coefficients equal zero

$\alpha = .05$

ANALYSIS OF VARIANCE

Source	DF	SS	MS	F
Regression	3	171.3	57.1	16.70
Error (unexplained)	11	37.6	3.41	

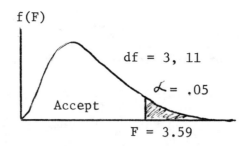

df = 3, 11
$\alpha = .05$
F = 3.59

Decision Rule: If: Calculated F > 3.59 reject H_o, otherwise do not reject H_o.

Since 16.70 > 3.59 we should conclude that the model is significant. However, while not all slope coefficients are thought to be zero, some may be.

The significance of the model is measured based upon the ability of the independent variables to collectively explain a significant portion of the variation in the dependent variable.

17-7 The percentage of variation in job rating explained by the three independent variables is 81.9 percent as measured by the coefficient of determination, R^2.

$$R^2 = \frac{SSR}{SST} = \frac{171.3}{208.9} = .819$$

17-9 Using the multiple regression model our best estimate of the job rating of this individual would be $\hat{Y}$, the point estimate from the regression model:

$$\hat{Y} = 21.4803 + 2.364(X_1) + 1.531(X_2) + 3.807(X_3)$$

$$\hat{Y} = 92.78$$

Our best estimate for this individual is a job rating of 92.78.

17-11 The 95% confidence interval for B_2, the coefficient for the variable, written score is found as follows:

$$b_2 \pm t_{\alpha/2\,(n-k-1)} S_{b_2}$$

$1.531 \pm 2.201\,(1.773)$

1.531 ± 3.902

$-2.371 \text{————} 5.433$

Our interpretation of this interval would be that of all possible intervals calculated from a sample of this size, 95% would include the true value of B_2. Therefore, we are confident that this interval includes the true B_2 value. Furthermore, because of the fact that this interval includes zero, we must conclude that the true slope may be zero.

Our response to this individual would be that it is our estimate that if he would increase the value of his written score by 1 point, holding the other variables constant, he can expect an average change in his job rating from -2.371 points to 5.433 points based upon the sample data.

17-13 We are asked to test the significance of each correlation coefficient. The general form of the null and alternative hypotheses is:

$$H_o\; \rho = 0.0$$

$$H_A\; \rho \neq 0.0$$

The appropriate test statistic is:

$$t = \frac{r}{\sqrt{(1-r^2)/(n-2)}}$$

The test is a t test with n-2 degrees of freedom. Thus, the tests we need to perform are:

For: (Y, X_1) (Y, X_2) (Y, X_3) (X_1X_2) (X_1X_3) (X_2X_3)

In all cases, the decision rule will be:

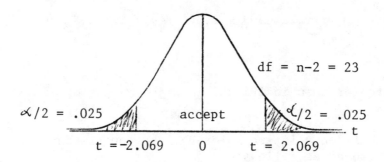

If calculated t > 2.069 or < -2.069 reject H_o, otherwise do not reject.

The t statistics are computed for each pair of variables:

(Y_1X_1)

$$t = \frac{-.4057}{\sqrt{(1-(.4057)^2)/(23)}} = -2.12 < -2.069 \therefore \text{reject } H_o$$

(Y_1X_2)

$$t = \frac{.4591}{\sqrt{(1-(.4591)^2)/(23)}} = 2.47 > 2.069 \therefore \text{reject } H_o$$

(Y_1X_3)

$$t = \frac{-.2444}{\sqrt{(1-(.2444)^2/23}} = -1.21 > -2.069 \therefore \text{do not reject}$$

(X_1X_2)

$$t = \frac{.0512}{\sqrt{(1-(.0512)^2)/23}} = .246 < 2.069 \therefore \text{do not reject}$$

(X_1X_3)

$$t = \frac{.5037}{\sqrt{(1-(.5037)^2)/23}} = 2.79 > 2.069 \therefore \text{ reject } H_o$$

(X_2X_3)

$$t = \frac{.2718}{\sqrt{(1-.2718^2)/23}} = 1.35 < 2.069 \therefore \text{ do not reject}$$

In summary, we have found that the following pairs of variables have correlations different from 0.0 as tested at the .05 alpha level:

$$(Y_1X_1) \quad (Y_1X_2) \quad (X_1X_3)$$

The other pairs do exhibit correlations which lead us to reject the null hypothesis.

You should note that each time a hypothesis test is performed, we run the risk of committing a Type I error. While the alpha for any one test is .05, the combined overall alpha for multiple tests is:

$$\text{alpha} \leq 1 - (1-\alpha)^r$$

where:

r = number of tests performed.

17-15 To be consistent with the alpha level used in problem 17-13, we will use α = .05. In practical applications, α should be set in accordance with the cost of committing a Type I error.

To test the significance of the model at step 1, we could proceed in a number of ways. For instance, we might hypothesize:

H_o : all slope coefficients = 0.0

H_A : not all slope coefficients = 0.0

α = .05

Then we can use the analysis of variance approach where computed F statistic = 6.14.

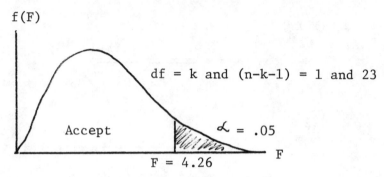

Since 6.14 > 4.26 we reject the null hypothesis and conclude the regression model is significant at Step 1.

A second approach is to test the following hypothesis:

$$H_o : B_1 = 0.0$$
$$H_A : B_1 \neq 0.0$$
$$\alpha = .05$$

Since at Step 1 of the model we have only one variable in the model, this hypothesis is equivalent to the previous one. The appropriate test statistic is:

$$t = \frac{b_1 - B_1}{S_{b_1}} = \frac{.00199 - 0}{.0008}$$

$$t = 2.478$$

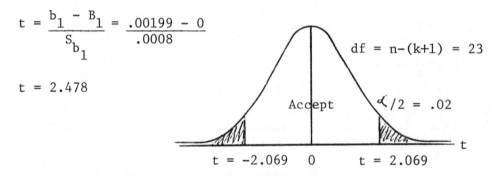

Since: $t = 2.478 > 2.069$ we reject the null hypothesis.

Both approaches for testing the significance of the model are equivalent.

17-17 At Step 2 of the model (Table D), variable X_1, AGE entered the model. At this point the regression model has the form:

$$\hat{Y} = 82.8875 - 1.31807\, X_1 + .00208\, X2$$

The inclusion of variable X1 had several effects. First, R^2 increased from .2107 at Step 1 of the model to .3955 at Step 2. The standard error of the estimate (SEE) declined form 36.3553 at Step 1 to 32.5313 at Step 2. Adjusted R^2 increased from .1763 to .3905 which indicates that the addition of variation explained has offset the cost, in terms of lost degrees of freedom, of bringing this variable into the model. Also, the coefficient on variable X2, Income, has changed slightly from .00199 to .00208.

17-19 The test of the individual significance of each independent variable in the regression model is performed as follows:

$H_o : B_1 = 0$ $\qquad$ $H_o : B_2 = 0$ $\qquad$ $H_o : B_3 = 0$

$H_A : B_1 \neq 0$ $\qquad$ $H_A : B_2 \neq 0$ $\qquad$ $H_A : B_3 \neq 0$

$\alpha = .05$ $\qquad\qquad$ $\alpha = .05$ $\qquad\qquad$ $\alpha = .05$

$$t = \frac{b_1 - B_1}{S_{b_1}} \qquad t = \frac{b_2 - B_2}{S_{b_2}} \qquad t = \frac{b_3 - B_3}{S_{b_3}}$$

$$t = \frac{-.97047 - 0}{.586042} = -1.6555 \qquad t = \frac{.00233 - 0}{.000745} = 3.132 \qquad t = \frac{-8.7233 - 0}{7.49549} = -1.163$$

The following t distribution is applicable for all three tests:

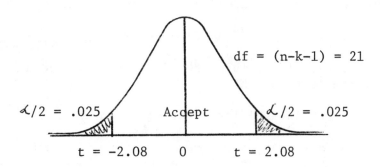

The decision rule in all cases is: If: $-2.08 \leq t \leq 2.08$ accept H_o, otherwise reject H_o.

Since only variable X2 has a regression coefficient which falls in the rejection region, we would conclude that given all three variables are included in the model, only X2 is significant in explaining the variation in the dependent variable.

$\qquad$ Now the test of the overall significance of the model is performed as follows:

$\qquad$ H_o : all slope coefficients = 0

$\qquad$ H_A : not all slope coefficients = 0

$\alpha = .05$

The appropriate test is the F test based upon the analysis of variance results in Table E. The calculated F = 5.33.

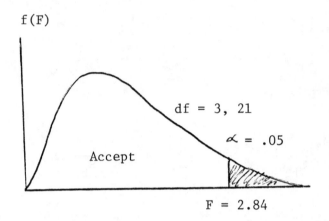

Since $5.33 > 2.84$, we reject H_o and conclude that <u>not all</u> slopes equal 0.0. Thus the overall model is said to be significant in explaining the variation in the dependent variable.

17-21 Possible points to consider in your report are:

1. Relatively low R^2 compared to 1.0

2. The relatively large standard error given the size of the dependent variable

3. The two insignificant variables in the model

4. The fact that there are many other variables which could have been included in the model

17-23 We have used a computer program to develop the regression model. The output of this program is shown as follows:

R^2	0.84479
Adjusted R^2	0.78271
Standard error	23849.68125

VARIABLES IN THE EQUATION

Variable	B	Std Error B	t
X1	175.8963	39.76977	4.422
X2	-1573.778	1995.85136	-.788
X3	1.591706	0.44446	3.581
X4	1613.747	625.02342	2.581
(Constant)	-125307.8		

Analysis of Variance	DF	Sum of Squares	Mean Square	F Ratio
Regression	4.	30960327043.09	7740081760.77379	
Error	10.	5688072956.90482	568807295.69048	13.607

Thus, the regression model is:

$$\hat{Y} = -125307.8 + 175.8963(X1) - 1573.778(X2) + 1.5917(X3) + 1613.747(X4)$$

17-25 The format for a confidence interval estimate for a regression slope coefficient is:

$$b_1 \pm t\, S_{b_1}$$

In this case, we want a 95% confidence interval estimate for each slope coefficient. The interval coefficient, t, is determined from the t table with n-k-1 = 10 degrees of freedom. Thus t = 2.228. The desired confidence intervals using the computer output in the solution to problem 23 are:

(X1) $175.89 \pm 2.228\,(39.769)$

175.89 ± 88.60

$87.29 \text{ ———— } 264.49$

(X2) $-1573.77 \pm 2.228\,(1995.85)$

-1573.77 ± 4446.75

$-6020.52 \text{ ———— } 2872.98$

(X3) $1.5917 \pm 2.228\ (.444)$

$1.5917 \pm .9892$

$.6025$ ——— 2.5809

(X4) $1613.747 \pm 2.228\ (625.02)$

1613.747 ± 1392.54

221.20 ——— 3006.29

In all cases, we interpret the interval to mean that we are 95% confident that the true regression slope coefficient has a value between the lower and upper limits.

17-27 The correlation coefficient between two variables is:

$$r = \frac{n \Sigma XY - \Sigma X \Sigma Y}{\sqrt{[n \Sigma x^2 - (\Sigma x)^2][n \Sigma y^2 - (\Sigma y)^2]}}$$

We have used a computer program to determine the following correlations:

(Y,X5) $r = .89579$

(Y,X6) $r = .66045$

17-29 We have employed a computer routine to develop the multiple regression model which includes all six independent variables. The output is shown as follows:

R^2 0.87252
Adjusted R^2 0.77691
Standard Error   24165.94185

VARIABLES IN THE EQUATION

Variable	B	Std Error B	t
X1	94.76854	86.04152	1.101
X2	-346.6237	2660.36945	-.130
X3	1.004265	0.73350	.1369
X4	1042.960	782.07397	1.334
X5	0.8365226	0.92714	.902

X6	596.0263	5749.30669	.104
(Constant)	-100834.5		

Analysis of Variance	DF	Sum of Squares	Mean Square	F Ratio
Regression	6.	31976458034.69720	5329409672.44953	
Residual	8.	4671941965.30279	583992745.66285	9.125

A regression model used for prediction purposes generally should contain only significant variables. Thus, to determine which variables to retain in the model, we should test the significance of the six variables using a t test with the following null and alternative hypothesis:

$$H_o : B_i = 0$$
$$H_A : B_i \neq 0$$
$$\alpha = .05$$

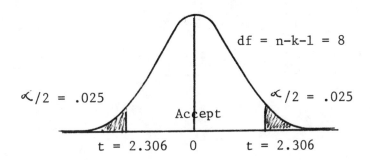

Thus, if any calculated $t = \dfrac{b_i - B_i}{S_{b_i}}$ falls in the rejection region, we will reject H_o and conclude that the variable is significant. Otherwise, we will accept H_o.

Using the output of the computer model we get:

Variable	Calculated T	Significant
X1	1.101	No
X2	.130	No
X3	.1369	No
X4	1.334	No
X5	.902	No
X6	.104	No

Thus, there are <u>no</u> significant variables in the model when all six variables are entered. A likely reason for this result is that we have six independent variables in the model and a sample size of only 15. Further, we suspect that X5 and X6 are correlated with each other and with the other independent variables resulting in some multicolinearity problems. We would suggest that a larger sample be selected and the model be rerun. Also, you might consider running a stepwise regression with a maximum of four variables in the model or that various combinations of four variables be tried in a series of regression models.

      As the model now stands, we would strongly recommend against using it for predictive purposes.

17-31 Certainly some degree of multicolinearity is likely to be present in the model. The correlation matrix indicates that some of the independent variables do overlap somewhat.

      The effects of multicolinearity are best seen in stepwise output where the size of the coefficients will change from step to step. Also, the signs on the regression coefficients may be different than the sign on the original correlation between the X and Y variables.

CHAPTER 18

TIME SERIES ANALYSIS

The previous 17 chapters have many times emphasized that business decision makers need information on which to base their decisions. While many sources of information exist, one very important source is historical data. If an organization has maintained historical records of such variables as sales, expenses, personnel turnover, and so forth, they can often look to these past data as a source of information about what the future may hold.

18-1 Time Series Components:

A Time series is a collection of observations of a particular variable measured over time. Any time series will contain one or more time series components. There are four potential components:

> Long Term Trend
> Seasonal Effect
> Cyclical Effect
> Random Variation

Long Term Trend:

The trend component is the long term increase or decrease in a variable over time. The greater the increase or decrease, the more pronounced the trend is. The following graph illustrates the trend in total sales for a small automotive supply business.

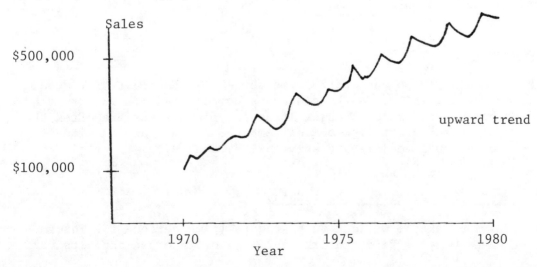

Seasonal Component:

The seasonal component represents those general changes in a time series which occur at the same time every year. Generally, the time series with a seasonal component contains measurements on a monthly basis with certain months during each year exhibiting higher values and other months having lower values for the response variable.

Many companies in the recreation, food, and entertainment business have seasonal sales levels. For example, the following graph shows the seasonality in the sales data for a hotel and golf course operation in Wisconsin.

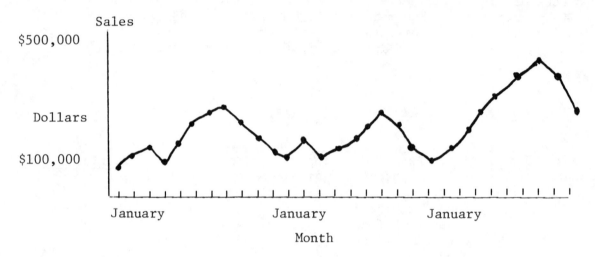

Cyclical Component:

Cyclical effects in a time series are represented by wavelike fluctuations around a long-term trend. These fluctuations are thought to often be the result of such factors as changes in interest rates, the money supply, unemployment, inventory levels, or any major governmental policy. Cyclical fluctuations tend to repeat themselves over the long run but will occur with differing frequencies and intensities.

Irregular Component:

The irregular component in the time series is that part of the variable which cannot be attributed to the preceding components. These irregularities can be both minor and major. Major irregular fluctuations are often the result of some "one time" event like the oil embargo of 1974 and the impact it had on the recreational vehicle industry in 1974 and 1975.

18-2   Analyzing the Variability of Past Data:

As decision makers, we look to past data (time series) for information. In most cases we will find that the variable of interest exhibits variation. The objective of the time series analysis techniques is to use the four time series components to explain the variability in the past data. The techniques presented in the text offer a basic introduction to time series analysis. More extensive treatments of the subject are contained in book references in the text.

18-3 Analyzing the Trend Component:

A primary means of analyzing the trend in a time series is to develop a regression model with a variable of interest as the dependent variable and the time unit being the independent variable. For example, if we expect that the trend is linear we would develop a regression model of the form:

$$\hat{y} = b_o = b_1 (t)$$

where:

$\hat{y}$ = estimated value of the time series variable

t = time period (1, 2, 3,. . . etc.)

The regression coefficient, b_1, provides an estimate for the average change in the time series variable for each period change in time. As with any regression model, a high R^2 will indicate that the trend is best modeled by the linear relationship between time and the variable of interest.

If the trend is not linear, we can develop other time series regression models. For instance, if the time series has exhibited an exponential growth pattern then we would try a model of the form:

$$\hat{y} = b_o + b_1 t + b_2 t^2$$

or

$$\hat{y} = b_o + b_1 t^2$$

18-4 Analyzing the Seasonal Component:

There are several ways of identifying and analyzing the seasonal component in a time series. The ratio to moving average method is commonly used. This method is based on the mutiplicative models:

$$y = (T)(S)(C)(I)$$

where:

y = value of the time series

T = trend value

S = seasonal value

C = cyclical value

I = irregular value

To analyze the seasonal component for monthly data, we begin by attempting to extract the T x C part of the model. To do this we first compute a <u>12 month moving</u> averages. We compute the moving average by averaging each set of 12 successive monthly time series values. Thus, if we have 24 months of data, we obtain 12 monthly average values. Note we do not get moving averages for the first 6 observations and the last 6 observations in the data.

The next step is to compute the <u>centered 12 month moving averages</u>. We do this by finding the mean of each successive pair of moving averages. Note, we lose one value on each end when we perform this step. However, we now have a centered moving average value corresponding to an actual time series value for all but the first seven and last seven cases.

Next, we divide the original data by the corresponding centered moving average to form the <u>ratio to moving average</u>. The ratios approximate the (S X I) or <u>seasonal index</u> factor in the mutiplicative model.

Then to eliminate the I or irregular component, we <u>normalize</u> the ratios by computing the average seasonal index for each month. We then multiply each average index by 1200 and divide each product by the sum of the seasonal indexes. The text illustrates this process in Table 18-3, and Table 18-4.

18-5 <u>Analyzing the Cyclical Component</u>:

The cyclical component is isolated by first removing the trend and seasonal compenents. One approach is to first develop a least squares trend line of the form:

$$\hat{y} = b_o + b_1 (t)$$

Then for each time period, (t), we compute the trend value, $\hat{y}$. Next we multiply the trend values by the seasonal indexes which were developed as discussed in section 18-4. This forms what we call the <u>statistical normal</u>.

Next, we isolate the combined cyclical and irregular components by dividing the actual observation of the time series variable by the corresponding statistical normal for each month. Generally the cyclical and irregular components are analyzed together as if they were one component.

18-6 <u>Analyzing the Irregular Component</u>:

The irregular component in a time series is considered to be unpredicatable fluctuations in the time series which cannot be attributed to the trend, seasonal, or cyclical components.

No statistical procedure exists for analyzing truly irregular or random fluctuations. Major irregular fluctuations can be smoothed out of the time series by taking a moving average of the data. The goal of smoothing is to eliminate irregular influences as much as possible, so that the true trend, seasonal and/or cyclical components can be recognized.

18-7   Index Numbers:

An index is constructed by dividing each value of the time series variable by a common base. The purpose of constructing index numbers is to provide a means of fairly comparing values measured over time.

The Problem of Different Bases

If the decision maker has reason to compare two or more sets of index numbers, he must make sure that the indexes have been developed with a common base. If not, a common base must be developed. As long as the data are unweighted we can accomplish this by dividing the existing index numbers for each index by the index number of a common base year. To transform this ratio to a percentage, we then multiply by 100.0.

18-8   Commonly Used Index Numbers:

There are several very prominent indexes which many decision makers rely on for a basis of comparing current economic activity with past activity. Some of the more common indexes are:

>Consumer Price Index
>Wholesale Price Index
>Industrial Production Index
>Dow-Jones Industrial Average
>Standard and Poor's 500

18-9   Forecasting:

While many decision makers have a need to analyze historical data and rely on the information from the past to help them make a wide range of current decisions, there is probably no greater use for time series analysis than that generated by the need for business and economic forecasts.

Forecasting involves making predictions about the future. This activity is required of almost every decision maker and includes forecasts of sales activity, inventory levels, machine down time, national economic performance, interest rates, and market size to name only a few areas.

Forecasting Methods:

There are two main categories of forecasting methods:
(1) judgmental
(2) statistical

Judgmental forecasts are forecasts made based upon a subjective assessment of the situation at hand. Oftentimes, experts are asked to evaluate a company's position with respect to a variable of interest and provide an estimate of the future activity of the company in that area. In other cases groups of people (often a combination of experts and non-experts) jointly evaluate the current situation and make forecasts.

Forecasting using the judgmental approach has the advantage of having no specific data requirements and is not tied directly to historical records or data. On the other hand, judgmental forecasting has the disadvantage of not being based on a sound statistical framework.

Statistical forecasting is a widely recognized discipline involving many statistical techniques ranging from some very basic techniques like moving averages and linear regression to very sophisticated techniques like the Box-Jenkins forecasting model. However, all statistical forecasting techniques share the common bond of being based upon the analysis of past data. Every statistical forecasting technique looks closely at the past. They attempt to determine the relationships between the variable of interest and the passage of time.

Statistical forecasting approaches include:

(1) Linear time series regression models

(2) Regression models with leading indicator variables

(3) Autoregressive models

(4) Exponential smoothing

Regardless of whether a forecast is made by judgmental means or through a complex statistical model, the ultimate objective is the same: to produce a "good" forecast. To test the model's forecasting ability, a recommended approach is to develop the model from a subset of the historical data available. Then use the resulting model to make forecasts for the periods for which the data were withheld. If the model performs adequately in these trials, the decision maker can feel somewhat assured that a similar model developed from the entire data will perform well under the real forecasting conditions.

However, there is no statistical forecasting model in existence which will provide accurate forecasts when the future suddenly looks very different from the past. However, some models can react more quickly to these changes.

18-10 Conclusions:

A time series may contain any one or more of the following components: trend, seasonal, cyclical, and irregular.

It is important for the decision maker to be able to look at the past and to determine how a variable of interest has behaved over time. This analysis may well provide an idea as to how this variable will behave in the future.

Forecasting is an activity all decision makers will be required to perform. The forecasting techniques, which are either judgmental or statistical in nature, have as their common objective a "good" forecast. The only way to truly evaluate a forecast is to wait and observe the actual outcome.

As with all managerial decisions, we remind you that the statistical analysis should only be considered as one source of information. The actual forecast produced by the decision maker should reflect a variety of information.

****
*   *
*   *
*   SOLUTIONS   *
*   *
****

18-1 For each of the industries listed, it is our objective to identify the particular time series component of most interest or importance. In most instances more than one component would be of interest.

  (a) Office Equipment Industry --
       <u>Long term trend</u> -- The office equipment industry no doubt expects growth over time.

       <u>Cyclical component</u> -- As the economy fluctuates, the demand for office equipment will cycle. In booming periods demand will be high, while in depressed periods demand may level out or even decline.

  (b) Recreational Vehicle Industry --
       <u>Seasonal component</u> -- It would seem logical that increases in demand would occur shortly before or during the summer months and taper off during the winter.

       <u>Irregular component</u> -- Unexpected jumps in gasoline prices has a substantial impact on the demand for R.V. products.

  (c) Tourism --
       <u>Cyclical component</u> -- Certainly the tourist industry is seasonal in nature. However, cyclical effects can be observed due to unusual weather patterns.

(d) Heavy Construction Equipment --

<u>Long term trend component</u> -- The heavy equipment industry, while moderately effected by seasonal variations is most concerned with long term growth trends.

<u>Cyclical component</u> -- This industry can also be impacted by cyclical variations due to factors such as government policies on construction, interest rates, etc.

(e) Banking --

<u>Long term trend component</u> -- While seasonal variations in the banking business can occur they are generally not as pronounced as for certain other industries. Generally the banking industry operates fairly consistently around a long term growth trend.

<u>Cyclical component</u> -- The banking industry is also impacted by cyclical factors such as government monetary policies. For example, in late 1979 and 1980, savings and loan institutions came upon hard times due to the Federal Reserve's monetary policy which forced up interest rates and depressed housing sales.

(f) Car Insurance --

<u>Trend component</u> -- While this industry is not totally protected from the seasonal and cyclical variations, the most significant component is the long term trend.

18-3 The monthly index number can be found using the ratio to moving average method.

Year/Month	Units Sold	12 Mo. Moving Total	12 Mo. Moving Ave	Ctred 13 Mo Moving Ave	Ratio to Moving Average (Percent)
Jan, 78	60				
Feb	70				
Mar	80				
April	90				
May	100				
June	110				
		1200	100.00	100.21	
July	120				119.75
		1205	100.42	100.62	
Aug	130				129.19
		1210	100.83	101.04	
Sept	140				138.56
		1215	101.25	101.46	
Oct	120				118.28
		1220	101.67	102.08	
Nov	100				97.96
		1230	102.50	102.92	
Dec	80				77.73
		1240	103.33	103.75	
Jan, 79	65				62.65
		1250	104.17	104.58	
Feb	75				71.71
		1260	105.00	105.21	
Mar	85				80.79
		1265	105.42	105.83	
April	95				89.76
		1275	106.25	106.88	
May	110				102.92
		1290	107.50	107.92	
June	120				111.20
		1300	108.33		
July	130				
Aug	140				
Sept	145				
Oct	130				
Nov	115				
Dec	90				

18-5. Earning figures with this much time difference are hard to compare. Certainly it is necessary to adjust these earnings data for the spending value in the associated time period.

18-7   The birth rate data are plotted on the following graph.

(a) For the total period, 1930 and 1975, there does not appear to be any identifiable trend. However, for the period since 1950, there is a definite downward trend in the birth rate per 1000 women in the United States. Using just the graph, there may be a minor long term trend in the 1930 to 1976 data. However, if long term is taken as 1950 to 1976 there is a definite, major, downward trend.

(b) There appears to be a long term cyclical component. However, the available data does not exhibit an entire cycle.

(c) No. If there is a cyclical component we would expect rates to increase again in the future. However, given this amount of data, it is possible that we have misread the cyclical component. It may not exist.

(d) The reasons people have children are likely based on variables other than the passage of time. Time is only a surrogate measure of these other variables and for these data appears to be a poor surrogate.

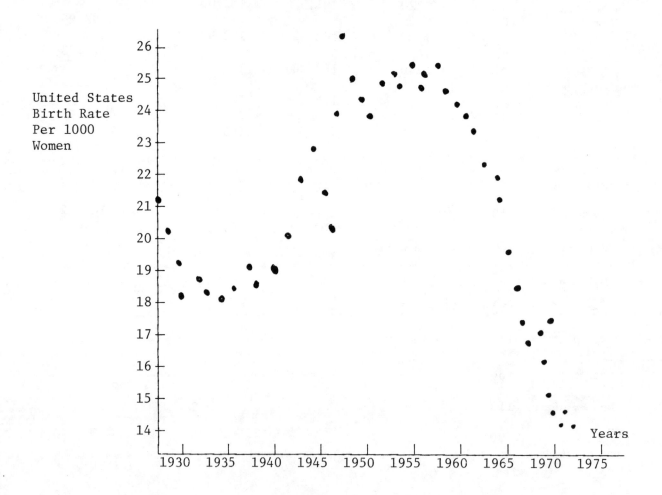

18-9   The data are plotted on the following graph.

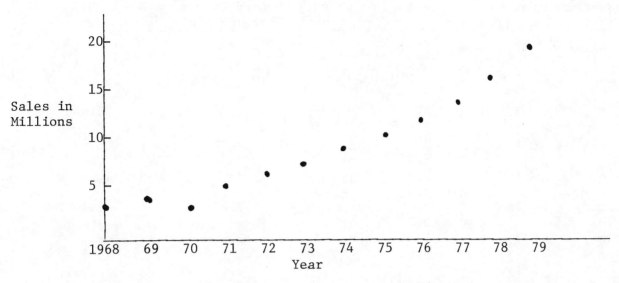

(b) There is a definite increasing trend in total sales over time. However, the trend is not perfectly linear. Regression analysis would likely help to identify and describe the trend in these data. We would explore models specified as follows:

$$\hat{y} = b_0 + b_1(\text{time})$$

$$\hat{y} = b_0 + b_1(\text{time}) + b_2(\text{time})^2$$

18-11   (a) The graph is shown below:

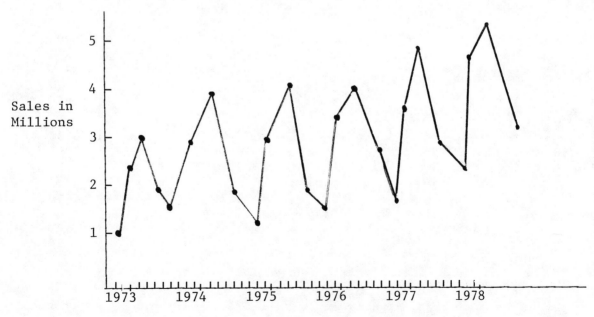

(b) and (c) are shown in the table below

Quarter/Year	Units Sold	4 Qtr. Moving Total	4 Qtr. Moving Ave.	Ctrd. 4 Qtr. Moving Ave.	Ratio to Moving Average
1st, 73	1.1				
2nd	2.4				
		8.1	2.02		
3rd	2.9			2.06	1.4061
		8.4	2.10		
4th	1.7			2.15	.7907
		8.8	2.20		
1st, 74	1.4			2.29	.6120
		9.5	2.37		
2nd	2.8			2.40	1.1667
		9.7	2.42		
3rd	3.6			2.41	1.4922
		9.6	2.40		
4th	1.9			2.41	.7876
		9.7	2.42		
1st, 75	1.3			2.46	.5279
		10.0	2.50		
2nd	2.9			2.51	1.1542
		10.1	2.52		
3rd	3.9			2.56	1.5220
		10.4	2.60		
4th	2.0			2.65	.7547
		10.8	2.70		
1st, 76	1.6			2.74	.5845
		11.1	2.77		
2nd	3.3			2.85	1.1579
		11.7	2.92		
3rd	4.2			2.96	1.4177
		12.0	3.00		
4th	2.6			3.06	.8490
		12.5	3.12		
1st, 77	1.9			3.20	.5938
		13.1	3.27		
2nd	3.8			3.34	1.1386
		13.6	3.40		
3rd	4.8			3.46	1.3863
		14.1	3.52		
4th	3.1			3.60	.8611
		14.7	3.67		
1st, 78	2.4			3.71	.6465
		15.0	3.74		
2nd	4.4			3.76	1.1694
			3.77		
3rd	5.1				
4th	3.2				

(d) The normalization factor is found below:

Quarter

Year	1	2	3	4
1973			1.4061	.7907
1974	.6120	1.1667	1.4922	.7876
1975	.5279	1.1542	1.5220	.7547
1976	.5845	1.1579	1.4177	.8490
1977	.5938	1.1386	1.3863	.8611
1978	.6465	1.1694		
Average Quarterly Index	1 .5929	2 1.1573	3 1.4448	4 .8086

The normalization factor will be 4 here since we are dealing with quarterly data.

$$\text{Normalization factor} = \frac{4}{\text{Average Quarterly Index}}$$

$$= \frac{4}{4.0036} = .9991$$

Noramalized Quarterly Index	1 .5923	2 1.1562	3 1.4434	4 .8078

18-13 Basic steps involved:

(1) Collect accurate and comparable sales data for as many past periods as possible.

(2) Identify potential explanatory variables such as advertizing, cattle volumes, national economic indicators, etc.

(3) Collect data for each of the explanatory variables.

(4) Perform time series analysis on the sales data attempting to extract trend, seasonal and cyclical components from the data.

(5) Develop a regression model using explanatory variables (step 2) as independent variables.

18-15   The index numbers will be constructed using the example of Table 18-8, with base equal to first quarter 1976.

Quarter and Year	Housing Starts	Index Numbers
1st, 76	110	100.0
2nd	112	101.8
3rd	107	97.3
4th	95	86.4
1st, 77	97	88.2
2nd	98	89.1
3rd	98	89.1
4th	105	95.5
1st, 78	113	102.7
2nd	123	111.8
3rd	117	106.4
4th	103	93.6
1st, 79	101	91.8
2nd	110	100.0
3rd	128	116.4
4th	136	123.6
1st, 80	130	118.2
2nd	122	110.9
3rd	132	120.0
4th	145	131.8

18-17   We will use Fall, 1976 as the base year in developing the index numbers. Note the base period is assigned an index number of 100. Index numbers for other periods are determined by dividing the number of students registering in the period in question by the base period students times 100. For example, for Winter, 1976:

$$\text{Index} = \frac{1128}{1325} (100)$$

$$\text{Index} = 85.1$$

Quarter and Year	Registration	Index Numbers
Fall, 76	1325	100.0
Winter	1128	85.1
Spring	814	61.4
Fall, 77	1267	95.6
Winter	1138	85.9
Spring	820	61.9
Fall, 78	1245	94.0
Winter	1074	81.1
Spring	753	56.8
Fall, 79	1130	85.3
Winter	1003	75.7
Spring	692	52.2
Fall, 80	1039	78.4
Winter	967	73.0
Spring	622	46.9

18-21. We have used a computer to develop the linear time series model of the form:

$$\hat{y} = b_0 + b_1 \text{(time)}$$

where:
$\hat{y}$ = monthly sales

time = month number (January, 1978 = 1)
(February, 1978 = 2), etc.

The resulting model (with November and December 1979 held out) is:

$$\hat{y} = 80.1299 + 2.103 \text{(time)}$$

The model has

$$R^2 = .271$$

$$S.E.E. = 22.91$$

$$S_{b_1} = .769$$

The model is significant at the alpha = .05 level with a calculated t determined as follows:

$$t = \frac{b_1 - B_1}{S_{b_1}} = \frac{2.03 - 0}{.769}$$

$$t = 2.64$$

This t value exceeds the critical t for alpha = .05 and 21 degrees of freedom = 2.080.

However, the percentage of variation explained is small and the S.E.E. is large relative to the size of the values of the dependent variable. To test the predictive ability, we forecast November and December 1979 sales and compare with the actual sales.

For November:
Forecast Sales = 80.1299 + 2.103 (23)
= 128.5
Actual November Sales = 115

For December:
Forecast Sales = 80.1299 + 2.103 (24)
= 130.6
Actual December Sales = 90

The forecasts are not too accurate. Given the rather low R^2 and high S.E.E. for the regression equation we would expect a poor forecast. We might point out that it appears that a linear time series model is not the most appropriate for these data. Therefore, it is somewhat unfair to attempt to predict two arbitrarily selected months' sales which happen to occur during the peak sales periods. Because the linear model is not sensitive to seasonality in the data, we might expect the under forecasts observed here.

18-23 Faced with a choice between the two models, we would pick the autoregressive form because it is more responsive to the seasonality in the data. However, in both cases, we would re-compute the model using all available data. That is, incorporate the previously held-out cases and determine the new regression coefficients before making a forecast for January, 1980 sales.

CHAPTER 19

INTRODUCTION TO DECISION ANALYSIS

Decision makers face the problem of making their decision in an uncertain environment. The previous 18 chapters have emphasized this concept and have introduced many "classical" statistical techniques which can be used to deal with the uncertainty.

However, the classical approach to decision making assumes that the only useful information about the population of interest is contained in a sample from that population. Chapters 19 and 20 introduce the subject of Decision Analysis or Bayesian decision theory. The techniques involved will present a means by which subjective (prior) information can be combined with sample information to provide information about the population of interest and hopefully lead to better decisions.

The decision maker can be faced with any of three decision environments; certainty, risk, and uncertainty.

19-1   Decision Making Under Certainty:

In those situations where the decision maker can identify all alternative options, and knows the outcome of each alternative, the decision environment is defined as certain.

The manager who is fortunate enough to operate in a certain environment faces only the task of identifying his or her alternatives. Then the decision to be made is the one which yields the optimum value, perhaps the highest profit or lowest cost.

Rarely does a decision maker get to operate in a certain environment.

19-2   Decision Making Under Risk:

The risk environment occurs when a manager is faced with an alternative with two or more possible outcomes and while this manager will not know which outcome will occur, he knows their associated probabilities.

For example, suppose a CPA firm is deciding whether to submit an audit proposal to a new company in the area. If they elect to submit a proposal, one of two outcomes will result; they will get the audit or they won't get the audit. Further, suppose that nine other firms have indicated that they will send in proposals too. If the selection is to be made on a random draw the CPA firm would know their probability of success (1/10).

Note, while the risk environment is more common than certainty, neither are as frequently occurring as the uncertainty environment.

19-3 Decision Making Under Uncertainty:

Uncertainty is like risk, except when a decision maker is faced with alternatives that have several outcomes, probabilities cannot be directly assigned to these outcomes. In this case, the manager must <u>assess</u> the probabilities such that they reflect his or her state of mind about the likelihood of each outcome occurring.

19-4 Subjective Probability Assessment:

When the probabilities connected with the decision outcomes are not known, they must be assessed subjectively.

A subjective probability assessment involves assigning a value between 0.0 and 1.0 to each outcome such that this assignment reflects your "state of mind" about the changes of the outcome occurring. The only restriction is that the sum of the probabilities for the possible outcomes of a single alternative must sum to 1.0.

19-5 Decision Making Criteria:

The basic premise of decisions analysis is:

<u>A "good" decision does not insure a "good" outcome.</u>

If we could always be assured a good outcome will result, decision making would be a lot easier. Since this isn't the case, the next best thing we can do is attempt to make "good" decisions. However, in order to determine whether we have made a good decision, we must have some decision criteria. Decision criteria are often divided into two classes: non-probabilistic and probabilistic.

Non-Probabilistic Criteria:

As the term implies, non-probabilistic criteria do not utilize the assessed probabilities in the decision process. Several non-probabilistic criteria are available:

<u>Maximin</u> - For each decision alternative, find the <u>minimum</u> possible payoff. Then pick the alternative with the greatest minimum.

<u>Maximax</u> - For each decision alternative, find the <u>maximum</u> possible payoff. Then pick the alternative which is with the greatest maximum.

Note, these two criteria are on the opposite extremes. The maximin is extremely conservative while the maximax is extremely optimistic. Both, however, suffer the shortcoming of not including the information contained in the decision makers probability assessment.

Probabilistic Criteria:

The fundamental probabilistic decision criterion is called the expected value criterion. This is based on the long run expected average occurrence.

The expected value criterion says that given a probability distribution of payoffs, select the option which yields the greatest expected payoff or minimum expected cost.

The expected value is:

$$E(X) = \sum X \, P(X)$$

Note, this expected value is nothing more than a weighted mean where the probabilities (P(x)) are the weights for each value of X.

You should note, the expected value is the long run average and may well never occur in any one case.

19-6   Decision Tree Analysis:

Decision Tree Analysis is a technique to aid the decision maker in dealing with complex decision problems involving several decisions, each of which may have several alternatives with several possible outcomes each.

The decision tree presents a network of the decision situation in a chronological order.

In decision analysis terminology a decision point is called an act and the possible outcomes are called events. Then in order to utilize decision tree analysis, we must organize the acts and events in proper sequence.

Once the chronological sequence is determined, the probabilities and cash flows are assigned to the decision tree. Then we work from the far right side of the decision tree working back to the left in a process called folding back the tree.

The expected values for each event are determined and then the alternative selected will be the one with highest expected payoff or lowest expected cost.

19-7   Conclusions:

This chapter has introduced the three potential decision environments and discussed two classes of decision criteria.

The text also illustrates through an example the value of utlizing decision tree analysis in the decision process.

The next chapter presents a formal means for combining subjective information and sample information to better assist the manager in making "good" decisions.

SOLUTIONS

19-1   Two examples which illustrate this concept which have recently come to our attention can serve as typical response you might provide.

The first involved a local county highway district which performed an extensive study of their computing needs. This involved a comprehensive systems analysis and design and a thorough analysis of hardware and software bids from over twenty vendors. The result of this careful analysis was the selection of a Portland, Oregon company which would provide both the hardware and custom software to meet the highway district's needs. By all indications, this firm "best" satisfied the decision criteria. However, while the hardware has proven to be quite satisfactory, the custom software has not lived up to expectations nor has it been delivered on schedule.

A second situation centered on an investment decision involving an international construction company. This ompany's stock by all financial standards should be selling between $10-$15 per share, yet as of December 11, the stock was selling for $29 per share and had been increasing steadily over the previous few months. A careful analysis of the company indicated that soon the stock must revert to a more "reasonable" price. The decision was to sell the stock short which essentially means that an investor will guarantee to sell stock he really doesn't own at a price (say $29 per share) in the hopes of being able to purchase the stock (to cover his sale) at a later date for a much lower price and thus turn a good profit. In this case, the stock continued to increase to $52 per share and eventually split 2 for 1 and the investor lost a great deal of money on his short sale.

19-3   The three basic decision environments are certainty, risk, and uncertainty. The certainty environment includes those decision situations for which the outcome is known with certainty. This environment is rarely present in business situations. The main consideration is that the decision model accurately reflects the situation. An example is the decision of whether to accept an hourly job at a fixed rate of pay. There is no uncertainty about the pay received for the hours work.

The risk environment represents decision situations in which the decision maker is faced with alternatives that, if selected, will result in one of several possible outcomes. The decision maker is aware of these potential outcomes and knows the probabilities associated with each outcome but cannot control the outcome.

The uncertainty environment is similar to the risk environment with the exception that the probabilities connected with the outcomes are not known. Rather they must be assessed by the decision maker so as to reflect the decision maker's state of mind relative to the likelihood of each outcome occurring.

19-5   This appears to be a decision making under certainty environment. All aspects of the decision are known. Pick-It-Up can accept or reject the option. However, over time, the number of customers may change and many aspects of risk and uncertainty will be involved in the business operations. To the extent that these are included in the initial analysis makes this a risk or uncertainty environment.

19-7   The number of units demanded is an event beyond the corporation's control. However, the decision is how many to build. We are given the following information:

$$\text{land cost} = \$100,000$$
$$\text{fixed building cost} - \$120,000$$
$$\text{variable building cost} - \$48,000$$
$$\text{selling price} = \$80,000$$
$$\text{salvage value} = \$21,000$$

The payoff functions become:

If demand is greater than or equal to supply:

Pay off = - $220,000 - $48,000 x units built + $80.000 x units sold

If supply is greater than demand:

Pay off = - $220,000 fixed cost - $48,000 x units built + $80,000 x units sold + $21,000 (units built - units sold)

Our first step in determining how many to build is to construct a Pay Off Table: (Assume the decision has already been made to build at least 10 units.)

```
                        Units Demanded
Units
Built      0         10         20         30         40         50
 10   -490,000    100,000    100,000    100,000    100,000    100,000
 20   -760,000   -170,000    420,000    420,000    420,000    420,000
 30 -1,030,000   -440,000    150,000    740,000    740,000    740,000
 40 -1,300,000   -710,000   -120,000    470,000  1,060,000  1,060,000
 50 -1,570,000   -980,000   -390,000    200,000    790,000  1,380,000
```

19-9   The maximax criterion selects the alternative which has the largest maximum gain. Using the payoff table for problem 19-7, we would find the highest payoff for each alternative level of production, assume it will occur, and then pick the alternative for which the value is greatest.

Units to Build	Highest Payoff
10	$ 100,000
20	420,000
30	740,000
40	1,060,000
50	1,380,000

Since $1,380,000 exceeds all other highest payoffs, Home-Sweet-Home should build 50 units using the maximax criterion.

19-11   The events beyond the Company's control are whether the beer is successful or not. Their decision is whether or not to market the new dark beer. The Payoff Table is:

	State of Nature Successful	Unsuccessful
Decision — Market	$1,000,000	-$625,000
Decision — Not to Market	-0-	-0-

19-13   There are now four possible states of nature as opposed to the two of problem 19-11. The Payoff table becomes:

	States of Nature Great	Good	Fair	Poor
Decision — Market	1,200,000	1,000,000	100,000	-625,000
Decision — Not to Market	-0-	-0-	-0-	-0-

19-15 The number of gallons demanded is an event beyond the owner's control. However, he must decide how many gallons to order. The payoff function differs depending on whether supply is greater or less than demand.

If demand is greater than or equal to supply:

Profit = ($6.25 - $2.50) x (demand)

If demand is less than supply:

Profit = ($6.25 - $2.50) (demand)
       -($2.50 - $1.00) (supply - demand)

Using these functions, the resulting payoff table is:

States of Nature
(Demands)

		500	1000	1500	2000
Amount Purchased	500	$1875	$1875	$1875	$1875
	1000	1125	3750	3750	3750
	1500	375	3000	5625	5625
	2000	-375	2250	4875	7500

19-17 The number of cakes demanded is beyond the manager's control, however, he can decide on the number of cakes to make. The profit function depends on whether demand is greater or less than supply.

If supply is less than or equal to demand:
Profit = (Selling Price - Variable Cost) (Demand)
Profit = ($8.00 - $2.25) (Demand)

If supply is greater than demand:
Profit = (Selling Price - Variable Cost) (Demand) - (Variable Costs - Salvage) (Supply - Demand)
Profit = ($8.00 - $2.25) (Demand) - ($2.25 - $1.25) (Supply - Demand)

Using these profit functions, the payoff table becomes:

                    States of Nature
                       (Demands)

		0	1	2	3	4	5
	0	-0-	-0-	-0-	-0-	-0-	-0-
Alterntiave	1	-1.00	5.75	5.75	5.75	5.75	5.75
Levels	2	-2.00	4.75	11.50	11.50	11.50	11.50
Of	3	-3.00	3.75	10.50	17.25	17.25	17.25
Production	4	-4.00	2.75	9.50	16.25	23.00	23.00
	5	-5.00	1.75	8.50	15.25	22.00	28.75

19-19 (a) The advantage of analyzing the decision problem of this nature using the decision tree analysis approach is that we are forced to determine the proper sequence of decisions and events. Also of importance, decision tree analysis forces us to consider future decisions and take these decisions into account before making the current decision. The appropriate tree for the Continental Automobile Agency decision is:

Part I

*These figures come from the tree chart on the next page.

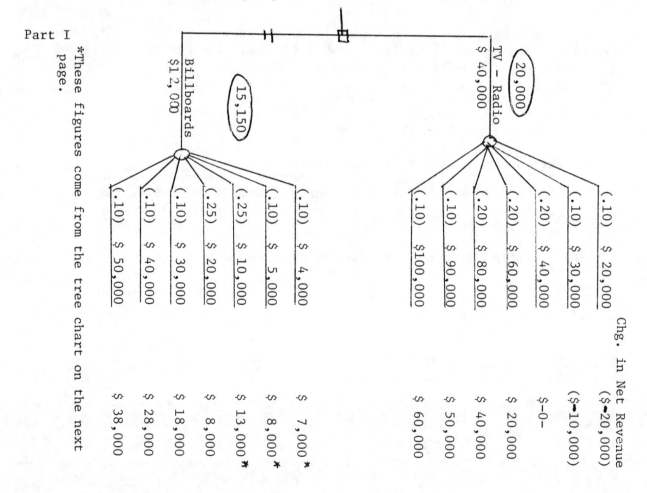

260

Part II

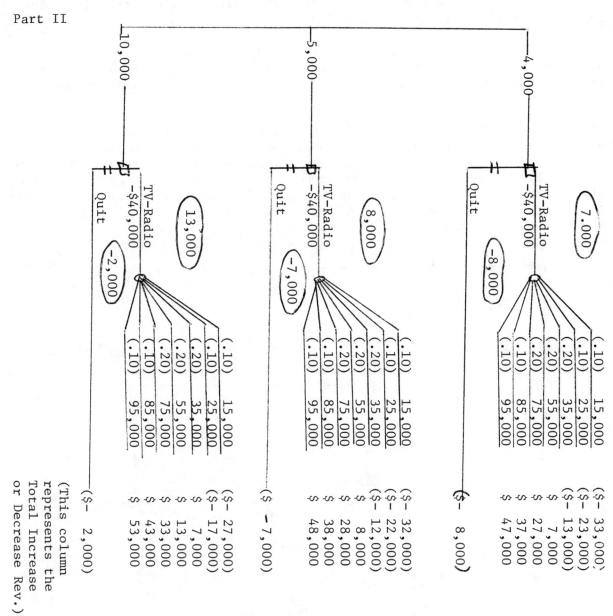

(b) The tree shown in part "a" also contains the cash flows and the probabilities. To determine the best decision using the expected value criterion we first determine the end values or final changes net revenues for each branch in the tree. (See decision tree). Next, we begin folding back the decision tree by first finding the expected values of the events on the right-most part of the decision tree.

For example, the expected value of the TV-Ratio Alternative in Part I of the decision tree is found as follows:

E(Net Revenue) = ($-20,000)(.10) + ($-10,000)(.10) + (0)(.20) +
($20,000)(.20) + ($40,000)(.20) + ($50,000)(.10) +
($60,000)(.10)

In a like manner the other expected values are determined as shown in the enclosed cirles on the decision tree.

When the tree is folded back to the far left, we find the expected value of the Billboard branch to be $15,150. Thus, since $15,150 is less than $20,000 the "best" decision is to select the TV-Radio campaign initially.

19-21   A decision tree approach can be used which will organize the decisions and events in the proper chronological sequence. The tree would look like the following. Keep in mind, the decision made in problem 19-20 was to accept the original offer from the grocery chain which had an expected value of $47,280. Note, we have assigned the probabilities (as determined in problem 19-20) and also the cash flows at the various points along the tree. The cash flows depend on the volume of apples and the price.

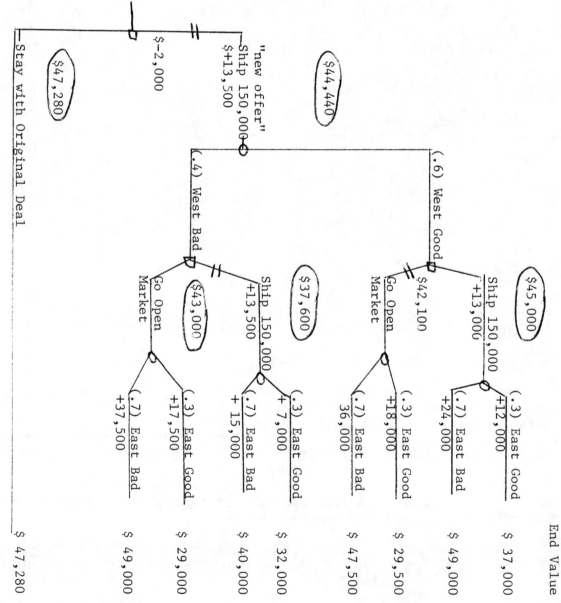

Now, we could fold back the decision tree from right to left. Note, the only branch that needs to be folded back is the "new offer" branch. We do this as follows:

Step 1:
$$E(value) = (\$37,000)(.3) + (\$49,000)(.7) = \$45,400$$
$$E(value) = (\$29,500)(.3) + (\$47,500)(.7) = \$42.100$$

Thus, if the west crop is good, the best decision is to follow through with another 150,000 pounds of the grocery chain at $.09 per pound since $45,400 is greater than $42,100.

Step 2:
$$E(value) = (\$32,000)(.3) + (\$40,000)(.7) = \$37,600$$
$$E(value) = (\$29,000)(.3) + (\$49,000)(.7) = \$43,000$$

Therefore, if the west crop is bad, sell all the remaining crop (250,000 pounds) on the open market since $43,000 is greater than $37,600. However, to arrive at the original decision we must do the following:

$$E(value) = (\$45,400)(.6) + (\$43,000)(.4) = \$44,440$$

Thus, the expected value of the "new offer" branch is $44,440. Now, to determine the optimal decision we compare the expected values of the two main branches. We find that $47,280 is greater than $44,440. Therefore, Balbado should not make the new offer. The markdown of $.01 per pound is not worth the chance to wait for information about the west crop size.

CHAPTER 20

BAYESIAN POSTERIOR ANALYSIS

The classical statistical techniques presented in chapters 1-18 for the most part considered the sample as the only source of information about a population. We discussed many statistical tests and techniques which utilized only sample information to provide information for the decision maker.

However, as we indicated in chapter 20, the subject of decisions analysis is based upon the premise that the decision maker may possess prior knowledge about the population of interest and that through subjective probability assessment he or she is capable of quantifying that knowledge. Given this assumption, the concept of Bayesian posterior analysis has been developed. In Bayesian posterior analysis, the decision maker has a formal way of combining his own prior knowledge along with new information from a sample or other sources. Together, the prior and new information offer the decision maker greater information on which to base the decision.

20-1 **Bayes' Rule Revisited:**

Bayes' rule offers the decision maker a special way of dealing with conditional probability. Formally, Bayes' rule is:

$$P(B \mid A) = \frac{P(B)P(A \mid B)}{P(B)P(A \mid B) + P(\bar{B})P(A \mid \bar{B})}$$

where:
- A = event A
- B = event B
- $\bar{B}$ = all events but B (B complement)

We see that Bayes' Rule is just an extension of the normal rule for conditional probability with the denominator expanded. The conditional probability rule is:

$$P(B \mid A) = \frac{P(B)P(A \mid B)}{P(A)}$$

Bayes' Rule simply substitutes in the denominator for $P(A)$, the sum of the probabilities of the ways event A can occur.

This same concept can be applied to managerial situations in which the decision maker wishes to combine some prior information with new information to arrive at a "posterior" level of knowledge. In this case we have the following:

P(B) = Prior probability

P(A | B) = Conditional probability of the sample information, A, given the value of B.

P(B)P(A | B) = Joint probability of prior and sample given prior.

P(A) = Sum of the joint probabilities over all levels of B.

In the next section we show how the Bayesian Posterior analysis works.

20-2 Bayesian Posterior Analysis:

Suppose a company which binds books for various publishers feels that the following probability distribution describes the defective binding distribution.

Proportion Defective	Probability (Prior)
.05	.60
.10	.30
.20	.10

For example, initially the company feels there is a 60 percent chance that .05 books will have a defect in the binding.

Now, suppose a sample of n = 20 books is selected at random from the warehouse and 3 books are found to be defective. The means by which this new sample information is incorporated along with the prior information is as follows:

Step 1: Determine the conditional probability of observing the sample results for each level of defective books. Assume the binomial distribution applies.

Proportion Defective	Prior	Conditional P(n=20, x1=3 | P)
.05	.60	.0596
.10	.30	.1901
.20	.10	.2054

Step 2: Next determine the joint probabilities by multiplying the priors by the conditionals and also sum the joint probabilities.

Proportion Defective	Prior	Conditional	Joint
.05	.60	.0596	.03576
.10	.30	.0901	.02703
.20	.10	.2054	.02054
			.08336

Step 3: Determine the posterior probabilities by dividing each joint probability by the sum of the joint proabilities.

Proportion Defective P	Prior	Conditional	Joint	Posterior
.05	.60	.0596	.3576	.4289
.10	.30	.0901	.02703	.3242
.20	.10	.2054	.02054	.2464
			.08336	

Now, the posterior probabilities reflect both the subjective prior information and the objective sample information. For instance, we see that the probability that there are .05 defective books has been reduced after including the sample information and the probability of .20 defectives has increased. This is consistent with what we would expect since the sample had 3/20 = .15 defectives.

If a second sample of n = 20 was selected, we could use the posteriors just computed as the priors and go through the process again to arrive at new posteriors. We could also combine the two samples and treat them as though they were one sample. The same exact posteriors will result.

The previous example utilized the binomial distribution in determining the conditional probabilities. However, any probability distribution could be used depending upon the sampling environment. Also, it is possible that both the prior information and the new information could be subjective. In any event, the process of obtaining the posterior probabilities is the same.

20-4 The Value of Information:

Added information generally comes with a price tag. It is important to know whether the information is worth its price.

When a decision maker operates in an uncertain environment it is likely that "poor" outcomes will result from "good" decisions. Thus, there is a cost of being uncertain. The value of sample information is based upon how successful this information is at reducing the cost of uncertainty.

Some important concepts are considered in determining the value of information:

Expected Value of Perfect Information:

This is the difference between the payoff that would result given <u>perfect information</u> and the payoff that results under uncertainty. This represents the upper limit on the value of any information.

Perfect Information:

Perfect information represents perfect knowledge about which outcome will result. However, perfect information provides knowledge only of the resulting outcome and not control over which will actually occur. Having perfect information is analogous to operating in a certain environment.

Cost of Uncertainty:

The cost of uncertainty is the difference between the payoff which would result if we had perfect information versus the payoff we expect without perfect information.

Value of Sample Information:

The difference between the cost of uncertainty before the sample information and the cost of uncertainty after the sample information is included in the decision.

20-5   Conclusions:

Decision makers often possess prior knowledge about the population of interest. To not recognize this knowledge in the decision process is a departure from the basic premise of this book; "good" decisions are those that are made with all available information considered in a logical manner.

Bayesian posterior analysis offers a systematic means for combining the subjective prior information with new information. We can also determine the value of the new information by measuring how much this information reduces our cost of uncertainty.

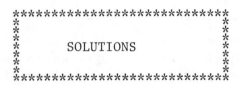

20-1   When is a prior probability not a prior probability? Answer - when it is a posterior probability.

While this line of reasoning may appear confusing, we think that in fact most probabilities which we think of as priors are actually posterior probabilities due to the idea that the probability assessment process is dynamic. That is, as decision makers, we constantly encounter information (both qualitative and quantitative) which is used to revise our thinking about the situation at hand. Therefore, there is a continuing process of probability revision taking place before the decision maker ever actually makes his or her assessment of the "prior" probability.

20-3   The reason large samples carry more weight in the revision process than do smaller samples is that, all other things being equal, large samples produce sampling distributions with less dispersion. This means that the larger sample will have a higher information content.

Another way of looking at this is that as the sample size increases there are more possible outcomes for the random variable $X1$. Thus, the conditional probability given any state of nature for any possible value of $X1$ will be smaller for a large sample size than for a small sample size. Therefore, when the conditional probability is multiplied by the prior, the value of the joint probability and likewise the posterior probability will more closely reflect the conditional probability than the prior.

20-5   The cost of assembling the toys will be different for the old and new machines

For the old machine:

Cost = \$2,800 + \$1.50 (Demand)

For the new machine:

Cost = \$11,000 + \$1.00 (Demand)

Using these cost functions, the payoff table becomes:

where: Payoff = \$3.25 (Demand) - Cost

Demand Levels

Old Machine Decisions	10,000	15,000	20,000	30,000	50,000
Old Machine	14,700	23,450	32,200	49,700	84,700
New Machine	11,500	22,750	34,000	56,500	101,500

20-7   The probabilities of problem 20-5 become the prior probabilities. The priors are multiplied by the conditional probabilities to determine a joint probability distribution. Then the posterior probabilities (reflecting both the prior information and the survey information) are determined as shown below.

Demand	Prior Probability	Conditional Probabilities	Joint Probabilities	Posterior Probabilities
10,000	.05	.05	.0025	.0057
15,000	.15	.10	.0150	.0343
20,000	.20	.30	.0600	.1371
30,000	.40	.50	.2000	.4571
50,000	.20	.80	.1600	.3657
			.4375	.9999*

*Difference from 1.00 due to rounding.

20-9   We will find the posterior probabilities using the table below:

Proportion Defective	Prior Probability	Conditional Probability	Joint Probability	Posterior Probability
$p = .05$	.85	$P(n=10, X1=4\|p=.05)$ .0010	.00085	$\frac{.00085}{.01405} = .06$
$p = .2$	.15	$P(n=10, X1=4\|p=.2)$ .0881	.0132 .01405	$\frac{.0132}{.01405} = .94$

Given this sample information, the probability the process is set up correctly drops from .85 to .06. This is due to the fact that if the machine is set up correctly there is little chance of finding 4 defectives in the sample of size 10.

20-11   The information is the same whether we take two samples ($n_1 = 10$ and $n_2 = 15$) or one sample of 25. Therefore, the posterior probabilties should be the same. This is shown in the following table:

Proportion Defective	Prior Probability	Conditional Probability	Joint Probability	Posterior Probability
$p = .05$	.85	$P(n=25, X_1=10\|p=.05)$ .00	.00000	.00
$p = .2$	.15	$P(n=25, X_1=10\|p=.2)$ .011	.00165 .00165	1.00

20-13   We begin by using the posteriors from problem 20-12 as the priors. Then because two samples of 25 each yield the exact same results as one sample of 50, we combine the last two samples to arrive at the new posteriors as follows.

Quality Level		Prior	Conditional Probability		Joint	Posterior
good	$p = .80$	.025	$P(n=50, X1=27\| p=.8)$	.0000	.0000	0
fair	$p = .60$	.939	$P(n=50, X1=27\| p=.6)$	.7780	.0730	.997
poor	$p = .30$	.036	$P(n=50, X1=27\| p=.3)$	.0067	.0002 .0732	.003 1.000

20-15   Perfect information is information which indicates beforehand which state of nature will occur. However, perfect information does not mean we can control which outcome will occur; rather we will simply know with certainty what the outcome will be.

20-17 King Construction has two choices:
(1) A sure profit of $125,000.

(2) A project with an expected value of:

$$E(\text{profit}) = \sum \text{profit} \times p(\text{profit})$$

$$= \$2,300,000 \times .3 + \$250,000 \times .3 + \$-1,200,000 \times .4$$

$$= \$285,000$$

King Construction should choose the development project since the expected payoff is greater than the alternative project.

20-19 King Construction would pay a maximum of $530,000 only if the consultants would guarantee they could specify the results of the development project beforehand

20-21 If we could know what Acme's percentage of defectives is in advance, we would buy from Windpro whenever Acme's defective rate is greater than 5 percent. The optimal decision table is shown below.

	Defective Rate	Number Defective Out of 100,000	Defective Cost/Best Company	Probability
Buy From Acme	.03	3,000	600,000	.2
	.04	4,000	800,000	.3
	.05	5,000	1,000,000	.2
Buy From Windpro	.06	5,000	1,005,000	.2
	.07	5,000	1,005,000	.1

Expected Breakage costs with perfect information would be:

$$E(\text{costs}) = (600,000 \times .2) + (800,000 \times .3) + (1,000,000 \times .2) + (1,005,000 \times .2) + (1,005,000 (.2 + .1))$$

$$= \$861,500$$

The expected value of perfect information is the difference between the expected cost under uncertainty and the expected cost given perfect information.

$$\text{EVPI} = 940,000 - 861,500$$

$$= \$78,500$$

20-23 To find the value of this sample information we will calculate a new EVPI given the sample of 100. The value of the sample will be found by taking EVPI of problem 20-21 minus the new EVPI. This will reflect how much the expected costs of uncertainty has been reduced by the sample. The new EVPI will be found using the following table.

	Defective Rate	Number Defective	Defective Cost	Probability
Buy from Acme	.03	3,000	600,000	.083
	.04	4,000	800,000	.263
	.05	5,000	1,000,000	.250
Buy from Windpro	.06	6,000	1,005,000	.276
	.07	7,000	1,005,000	.128

Now:

E(costs) = ($600,000 x .083) + ($800,000 x .263) + ($1,000,000 x .25) + ($1,005,000 x .276) + ($1,005,000 x .128)

= $916,220

Before we took the sample our cost of uncertainty was $78,500. Our cost of uncertainty (EVPI) after the sample is found as follows:

EVPI = (Expected Cost Uncertainty) - (Expected Cost PI)

$1,005,000 - $916,220

EVPI = $88,780

Thus, the sample has actually increased our uncertainty.

Value of sample = $EVPI_{before}$ - $EVPI_{after}$

= $78,500 - $88,780

= $-10,280

Thus, the sample has actually increased our uncertainty and in that respect has negative value.